The Latin American Studies Book Series

The Latin American Studies Book Series promotes quality scientific research focusing on Latin American countries. The series accepts disciplinary and interdisciplinary titles related to geographical, environmental, cultural, economic, political, urban and health-related research dedicated to Latin America. The series publishes comprehensive monographs, edited volumes and textbooks refereed by a region or country expert specialized in Latin American studies.

The series aims to raise the profile of Latin American studies, showcasing important works developed focusing on the region. It is aimed at researchers, students, and everyone interested in Latin American topics.

Submit a proposal: Proposals for the series will be considered by the Series Advisory Board. A book proposal form can be obtained from the Publisher, Andrea Brody-Barre (andrea.brody-barre@springer.com).

Mohammad Tarikul Islam · Md Belal Hossen

Economic Diplomacy

Reshaping Bangladesh-Latin American Diplomatic Relations

Springer

Mohammad Tarikul Islam
Department of Government and Politics
Jahangirnagar University
Dhaka, Bangladesh

Md Belal Hossen
Department of Government and Politics
Jahangirnagar University
Dhaka, Bangladesh

ISSN 2366-3421 ISSN 2366-343X (electronic)
The Latin American Studies Book Series
ISBN 978-3-031-81361-0 ISBN 978-3-031-81362-7 (eBook)
https://doi.org/10.1007/978-3-031-81362-7

This Springer imprint is published by the registered company Springer Nature Switzerland AG
The registered company address is: Gewerbestrasse 11, 6330 Cham, Switzerland

To our parents, who nourished the knowledge that grew inside of us so that we could develop into decent people.

Foreword

I am happy to hear that Springer Nature is very interested in publishing the new book that Prof. Mohammad Tarikul Islam and Md Belal Hossen wrote on the role of economic diplomacy in strengthening bilateral relations between Bangladesh and Latin American countries. I really appreciate their combined efforts. Given their originality, it includes a variety of issues that could not have been included in previous textbooks. These include, but are not restricted to, new areas of mutual cooperation to promote economic interests and economic diplomacy between Bangladesh and Latin American countries. In these joint efforts, there has been much discussion on how Bangladesh, having moved from the LDC to the developing countries, might deepen its economic ties with the Latin American region.

The book is an invaluable addition to the contemporary scholarships and is genuinely thorough in its covering. The book demonstrates rather well the possibilities that the bilateral ties between Bangladesh and Latin American nations have to promote economic expansion and ease commerce. Concurrently, the possibility of collaborating with each other in the field of international politics and handling unfavorable situations worldwide is also covered.

I am confident in stating that this book will serve as a helpful teaching tool and a helpful reference for further research.

Professor Michael Keith
Director of the PEAK Urban Research Programme
Former Director of the Centre on Migration
Policy and Society (COMPAS)
University of Oxford
Oxford, UK

Preface

Even when there are notable differences in the civilization, culture, language, and population of two countries—including those that are geographically apart—international relations are usually formed on the basis of economic and trade factors. When considering the relationship between Bangladesh and Latin American nations, it is clear that, despite possible differences in social and cultural harmony, there is mutual economic dependency. Thus, employing an advanced economic diplomacy approach may result in advantageous consequences for the development of Bangladesh's bilateral ties with its neighbors. Bangladesh trades internationally with almost all countries. Nonetheless, a large percentage of global trade suggests a strong reliance on specific nations. To transition from an emerging economy to a high-income, financially independent country, trade dependency must be reduced. However, it is unfortunate to note that Bangladesh has not kept up with other nations in forging bilateral ties or expanding its economy, even though it gained its independence over fifty years ago. As a result, Bangladesh continues to rely mostly on imports from China and India and exports to the United States and Europe. Because of this economic interconnectedness, the superpowers have the ability to take the initiative in Bangladesh's internal affairs. The people of Bangladesh are great fans of South American football, which is why they are considering using sports diplomacy to strengthen ties between their two countries. At different times, the respective governments of Uruguay, Brazil, and Argentina have all expressed their gratitude in public for Bangladesh's support. Bengalis have always had a deep connection to and interest in Latin football.

Rather than focusing on Africa, Eastern Europe, or the Middle East, Bangladeshi entrepreneurs are increasingly looking across the Atlantic at Latin America. As a result, Bangladesh's trade with this specific region has significantly increased. Nonetheless, there hasn't been a noticeable improvement in bilateral ties with the 13 nations listed below. Argentina has been surpassed recently in the level of football devotion among Bangladeshis. The Dhaka embassy has reopened after a 45-year hiatus. Brazil's sole previous connection to Dhaka was through its diplomatic mission. It's also important to note that Bangladesh only has diplomatic posts in two Latin American cities at the moment. Brasilia, the capital of Brazil, is home to one, while Mexico City is the location of the other. Bangladesh can enhance its focus on non-traditional trading partners such as Latin American and African nations by implementing measures to reduce its reliance on traditional trading partners. In this setting, the establishment and expansion of an all-encompassing framework for economic diplomacy are necessary. The notion for a thorough examination of Bangladesh's economic diplomacy with Latin American nations originated for the reasons outlined above. Multinational corporations, experts, and diplomats all acknowledge the importance of this work. It helps us feel more objective, responsible, and competent when we express ideas, viewpoints, and novel economic diplomacy initiatives in the document. Academicians interested in Bangladesh's non-traditional trading partners and economic diplomacy will continue to use this book as a reference. In addition, there is a great deal of hope that the research-based book will help to fortify Bangladesh's trade and economic diplomacy in the Latin American area.

Savar, Dhaka, Bangladesh

Md Belal Hossen
Mohammad Tarikul Islam

Acknowledgements

In doing this research to comprehend the dynamics and implications of economic diplomacy to influence bilateral relations between Bangladesh and Latin American countries, we have had the honor of getting substantial experience from a distinguished and highly competent academic. We might draw inspiration from this phenomenon to continue achieving academic success in the years to come. In addition, we received priceless guidance and support from distinguished department faculty members during the study term, which has had a significant impact on the development and refinement of our intellectual worldview. This has been used several times during the investigation. Working with diplomatic staff was necessary to collect and analyze data for this study on economic diplomacy and Latin America. Regarding this matter, we received significant support from several officials of Latin America Wing of the Ministry of Foreign Affairs, Government of Bangladesh, including Ambassadors.

Additionally, Muhammad Said of Oxford University contributed pertinent, up-to-date knowledge to enhance the literature review section. We acknowledge the inspiration provided by Prof. Dr. Md. Shamsul Alam of the Department of Government and Politics at Jahangirnagar University and Professor Bashir Ahmed, the Dean of the Faculty of Social Sciences of Jahangirnagar. We should be grateful to Mr. Zaki Faisal, the UNDP program Officer for his cooperation with scholarly inputs. Finally, but just as worthy of our gratitude is Shihab Uddin, a master's student working under co-author (Prof. Dr. Mohammad Tarikul Islam), who gathers news from the newspapers to aid in our academic project.

Md Belal Hossen
Mohammad Tarikul Islam

Contents

About the Authors

Dr. Mohammad Tarikul Islam is a Professor of Government and Politics at the Jahangirnagar University in Bangladesh. Prior to joining the University in 2014, Dr. Islam worked with the United Nations Development Program Bangladesh for a period of seven years in research, knowledge management, and project management capacities. Professor Islam acts as a Research Guide to different courses (Senior Staff Course—SSC and Advanced Course for Administration and Development—ACAD) at Bangladesh Public Administration Training Centre since 2019. Dr. Islam is a regular contributor to the South Asia Blog of the London School of Economics and Political Science (LSE), Oxford Political Review, SOAS Blog and the Institute of South Asian Studies, National University of Singapore. His edited volume *Human Security, Peace and Development: South Asian Perspective* has gained significant scholarly reviews. Professor Islam has also three major published publications to his credit. Among them are *Local Government in Bangladesh: Contemporary Issues and Challenges* from Routledge, *Disaster, Governance and Development: Bangladesh Perspective* published by Springer Nature, and an edited book *COVID-19 in South Asia: Its Impact on Society, Economics, and Politics* published by Routledge. He has given lectures and presented papers at seminars and conferences in Cambridge, LSE, SOAS, Oxford, Harvard, and many universities in China, India, Indonesia, Sri Lanka, Bhutan, and Nepal. Professor Islam is perhaps the rare academic with a Ph.D. from Bangladesh (Jahangirnagar) who has been affiliated with the University of Oxford, Harvard University, the University of Cambridge, and the School of Oriental and African Studies (SOAS) in different prestigious capacities (Visiting Scholar, Post-Doctoral Fellow, Guest Faculty and Visiting Research Fellow). His areas of interest in teaching and research include local government, crisis management, human security, climate diplomacy, sustainable development, project management, and forced migration.

Md Belal Hossen comes with a solid background in the Department of Government and Politics at Jahangirnagar University. He has been actively involved in Journalism since 2018. During that period, he provided event and investigative reports, authored features, and generated content for many notable national media outlets.

He held the position of president of the Jahangirnagar University Journalists' Association. During his time at the university, he also worked for Ajker Patrika, The Daily Campus as a university correspondent. Presently, he is working as a Reporter at The Business Standard. In addition, he is also affiliated with the Nepal Institute for International Cooperation and Engagement (NIICE) as a research intern. He is interested in economic diplomacy, trade, military, and security concerns, as well as local government and development.

Abbreviations

ADB	Asian Development Bank
APTA	The Asia-Pacific Trade Agreement
BAPI	Bangladesh Association of Pharmaceutical Industries
BASIS	Bangladesh Association of Software and Information Services
BB	Bangladesh Bank
BBS	Bangladesh Bureau of Statistics
BEZA	Bangladesh Economic Zones Authority
BGMEA	Bangladesh Garment Manufacturers and Exporters Association
BIDA	Bangladesh Investment Development Authority
BIMSTEC	Bay of Bengal Initiative for Multi-Sectoral Technical and Economic Cooperation
BIT	Bilateral Investment Treaties
BOI	Board of Investment
BOP	Balance of Payments
BPM	Balance of Payments and International Position Manual
CARICOM	Central American Common Market
D-8	Developing-8
EBA	European Union's Everything But Arms
ECLAC	United Nations Economic Commission for Latin America and the Caribbean
ECOSOC	Economic and Social Council
ED	Economic Diplomacy
EEZ	Exclusive Economic Zone
EPZ	Export Processing Zones
EU	European Union
FAO	Food and Agriculture Organization
FDI	Foreign Direct Investment
FOC	Foreign Office Consultations
FTA	Free Trade Agreements
FTS	Foreign Trade Statistics
GATS	General Agreement on Trade in Services

GDP	Gross Domestic Product
IDB	Islamic Development Bank
IMF	International Monetary Fund
IMF	World Bank and International Monetary Fund
IT	Information Technology
LA	Latin America
LA-BCCI	Latin America-Bangladesh Chamber of Commerce and Industry
LAC	Latin American Countries
LDC	Least Developed Country
MERCOSUR	The Southern Common Market (Spanish initials)
MoFA	Ministry of Foreign Affairs
MoU	Memorandum of Understandings
NBR	National Board of Revenue
ODA	Official Development Assistance
OIC	Organization of Islamic Cooperation
PA	Pacific Alliance
PTA	Preferential Trade Agreement
RMG	Ready Made garments
SAARC	South Asian Association for Regional Cooperation
SDG	Sustainable Development Goals
SEZ	Special Economic Zone
TI	Transparency International
UN	United Nations
UNCTAD	United Nations Conference on Trade and Development
UNDP	United Nations Development Program
UNFCCC	United Nations Framework Convention on Climate Change
UNGA	United Nations General Assembly
WEC	World Economic Forum
WTO	World Trade Organization

List of Figures

List of Tables

List of Maps

Chapter 1
Introduction

Abstract In an era defined by globalization and multipolarity, economic diplomacy has become a cornerstone for nations seeking to diversify their trade networks and establish strategic international partnerships. For Bangladesh, a rising economic power in South Asia, Latin America represents an uncharted yet promising frontier. Despite geographical distance, cultural differences, and historically limited engagements, this study explores the transformative potential of economic diplomacy in reshaping Bangladesh's diplomatic and economic relations with Latin American nations. Latin America, encompassing 13 independent nations and marked by a history of resource-driven growth, industrial expansion, and integration efforts like MERCOSUR, offers substantial opportunities. Its markets display growing demand for Bangladeshi exports such as ready-made garments, pharmaceuticals, IT services, and leather goods. Conversely, the region's rich agricultural outputs, mineral resources, and industrial products provide Bangladesh with access to cost-effective imports. However, despite the visible economic synergies, trade relations remain nascent, with Bangladesh lagging behind regional peers like India and China. For example, India's trade volume with Latin America reached $50 billion in 2022, while Bangladesh's trade remains modest at $621 million during the fiscal year 2022–2023, although showing promising growth. This research highlights the pivotal role of economic diplomacy in overcoming entrenched barriers to bilateral engagement. Key obstacles include logistical complexities, the absence of direct shipping routes, cultural and linguistic differences, and diverse regulatory frameworks. Additionally, Bangladesh's diplomatic footprint in Latin America remains minimal, with missions only in Mexico and Brazil, limiting its ability to foster sustained economic and political ties. Yet, recent developments, such as Argentina reopening its embassy in Dhaka after 45 years, signal a renewed momentum in Bangladesh–Latin American relations. Drawing on a mixed-method approach that combines secondary data analysis with expert interviews, this study examines the potential for Bangladesh to emulate successful economic diplomacy practices demonstrated by India and China in the region. It delves into strategies for leveraging technological advancements, such as improved maritime communication and digital diplomacy, to bridge geographical divides. Moreover, it explores the impact of global geopolitical shifts—such as the U.S.-China rivalry—on creating new trade opportunities for Bangladesh in Latin

M. T. Islam and M. B. Hossen, *Economic Diplomacy*, The Latin American Studies Book Series, https://doi.org/10.1007/978-3-031-81362-7_1

America. The research posits that strategic economic diplomacy can unlock vast potential in bilateral trade. It advocates for sector-specific collaborations in areas like textiles, agro-industrial goods, technology transfer, and sustainable energy. Moreover, integrating Bangladesh into regional trade agreements such as MERCOSUR could enhance its access to Latin American markets. In this context, the study underscores the importance of institutional capacity-building, including the establishment of dedicated trade missions and enhanced training for diplomats and trade negotiators. By fostering robust economic ties, both regions stand to benefit from diversified trade portfolios, enhanced export competitiveness, and sustainable development. Latin America can capitalize on Bangladesh's manufacturing expertise and burgeoning economy, while Bangladesh gains access to new markets and critical resources, reducing dependence on traditional trade partners. Beyond economic gains, such engagement promotes geopolitical stability and mutual understanding, positioning Bangladesh as a significant player in global trade networks. This study contributes to the broader discourse on international economic diplomacy by identifying actionable strategies to strengthen Bangladesh–Latin American relations. It calls for a pragmatic, multi-pronged approach that includes policy innovation, business-to-business engagement, and people-to-people exchanges. Ultimately, this research envisions a future where economic diplomacy catalyzes a dynamic and mutually beneficial partnership between Bangladesh and Latin America, driving shared prosperity and reinforcing their positions in the global economic landscape.

Keywords Trade relations · LAC · MERCOSUR · Market economy · Economic diplomacy

1.1 Background

Though the Latin American region might not have reached the same level of development as Europe or America, it does not lag behind Asia in terms of progress. The geographical area including the Maya civilization has exhibited considerable prosperity in the domains of agriculture, mining, and industry from antiquity. Because of this, the region has been under the influence of Europeans' desire for colonial expansion since the fourteenth century. The contemporary region of Latin or South America encompasses a total of thirteen independent nation-states situated within the geographical expanse between the Atlantic and Pacific oceans. The economic institutions in Latin America underwent a steady evolution throughout the nineteenth century, influenced by the fluctuating dynamics between democracy and militarization. In recent times, there has been a substantial increase in the magnitude of global trade, primarily attributed to the widespread establishment of liberal democratic systems across numerous nations in the region. An area of new promise has emerged for rising countries such as Bangladesh.

The concept of 'economic diplomacy' emerged throughout the latter half of the nineteenth century. Despite the absence of a universally agreed-upon definition within academic discourse, scholars have approached the subject matter from several perspectives, leading to a range of interpretations and analyses. The practice of economic diplomacy often pertains to the endeavors of a nation-state to achieve its economic goals and objectives through the identification and implementation of appropriate means and policies. Hence, the implementation of measures aimed at promoting economic growth through the prioritization of state interests in areas such as bilateral trade, import-export policies, economic agreements, tax and customs determination, international market policies, and foreign investment aligns with the principles and objectives of economic diplomacy. International relations are typically established based on economic and trade considerations, even in the presence of significant disparities in civilization, culture, language, and populace between two countries, including those geographically distant from each other. In the context of the relationship between Bangladesh and Latin American countries, it is evident that while social and cultural concord may be lacking, there exists a mutual economic interdependence. Consequently, the implementation of a sophisticated economic diplomacy strategy can potentially yield favorable outcomes for the advancement of Bangladesh's bilateral relations with neighboring nations. As has been observed in the countries of India, China, and Pakistan. On the other hand, bilateral relations are direct channels of communication between two sovereign governments on a variety of topics, such as politics, the economics, and culture, and they are established on the basis of shared understanding on both sides. The nature of this bilateral partnership is centered on the domestic policies of the state. The bilateral relationship undergoes modifications in response to transitions in government or changing circumstances. In the contemporary period of intensified global competition, the significance of international relationships has become increasingly multifaceted. The enhancement of bilateral relations has the potential to foster economic growth and facilitate trade between two nations. Simultaneously, the potential for mutual cooperation in the realm of international politics and the management of global adverse circumstances emerges.

Bangladesh's economy may benefit from a new source of low-priced but high-quality food stuff, cooking oil, sporting goods, mineral, and chemical raw materials: Latin America. In contrast, the Latin American market has shown a growing interest in Bangladeshi products, including textiles, medicine, leather goods, IT services, motor vehicles, and agricultural items, seeing the significant opportunities they present. In contrast, our neighbor India has already established significant public-private commercial ties in Latin America. In 2022, the trade between India and the area achieved a record-breaking milestone, amounting to $50 billion. In contrast, China's trade volume amounted to approximately 400 billion dollars. In this context, Bangladesh lags significantly behind its neighboring countries.

The Latin American region was not foremost on India's economic or political agenda, and by the late twentieth century, India was still a distant, minor market for Latin America. On the contrary, the twenty-first century has witnessed a sudden and profound transformation. Currently, India holds a significant position as one of the principal trading partners of Latin America, and reciprocally, Latin America also holds a comparable status in India's trade relations. If the region were considered as an independent nation, it would rank as the sixth largest trading partner for India. The bilateral relationship between India and Latin America is predominantly influenced by trade and economic factors. In addition to being a component of the 'Goldilocks Zone' for Indian business, Latin America will continue to be important to India's aspirations to become a major world force. Numerous Indian businesses are beginning to understand that they need to have a respectable presence in Latin America in order to be considered genuinely 'global.'

India has successfully cultivated bilateral relationships with all countries in Latin America through the execution of economic diplomacy. However, the Bangladeshi government and private sector have not yet reached the desired level of trade relations with this region. In essence, the closer establishment and development of bilateral relations with these countries would be facilitated if Bangladesh actively demonstrates its strengths in diplomatic economy. Based on the data provided by the Bangladesh Export Promotion Bureau, there was a notable 20.51% surge in Bangladesh's overall merchandise exports to Latin American nations, reaching a total value of 621.27 million dollars during the fiscal year 2022–23. The amount was $515.50 million during the fiscal year 2021–2022. With the recent growth of Bangladesh's economy, business organizations within the country are actively seeking opportunities to explore new markets across various regions globally.

Consequently, instead of focusing on the Middle East, Eastern Europe, or Africa, the Bangladeshi business community is increasingly looking across the Atlantic at Latin America. As a result, Bangladesh's trade with this specific region has significantly increased. Nonetheless, there hasn't been a notable improvement in bilateral ties with the 13 nations. Recently, Bangladeshis have surpassed Argentines in their passion for football. The embassy in Dhaka has been reopened after a period of 45 years. Prior to this, the only presence of Brazil in Dhaka was through its diplomatic post. Additionally, it is worth noting that Bangladesh currently maintains diplomatic missions solely in two cities within the Latin American region. One is in Mexico City, and the other is in Brazil's capital city of Brasilia.

In 2020, Argentina's exports to Bangladesh amounted to a value of $450 million, which subsequently experienced a notable growth to $791 million during the fiscal year 2021–2022. Argentina's primary exports to Bangladesh were wheat, soybean meal, and soybean oil. In light of the current global challenges pertaining to escalating prices and worldwide inflation, Argentina emerges as a prospective alternative market for Bangladesh. In contrast, Bangladesh exported items to Argentina with a total value of $17.3 million in the year 2020. The exports from Bangladesh to Argentina have shown a consistent annual growth rate of 6.46 percent over the course of the last 25 years. This growth has been observed starting from a relatively modest initial value of $3.61 million in the year 1995. The primary exports from Bangladesh to Argentina consisted of knitted sweaters, knitted t-shirts, and non-knit men's outfits. The ready-made garment industry accounted for about 88 percent of the total exports during the fiscal year 2021–2022. As a result, it is anticipated that the volume of exports and imports will increase at a nearly threefold pace following the development of bilateral connections between Bangladesh and Argentina.

Economic relations can have a major impact on the development and expansion of bilateral relations between Bangladesh and Latin American nations, according to analysts because Bangladesh and these countries do not share any historical or cultural traits. For this reason, the development of strong economic diplomacy could lead to improved business relationships, which in turn could foster bilateral dialogue and relationship-building. The current status of relations between Bangladesh and Brazil is an excellent example of this phenomenon. Brazil's economic interdependence with Bangladesh is predominantly reliant on the importation of goods. In contrast to its relations with other 12 countries in the Latin area, Bangladesh maintains favorable bilateral relations with this particular country. Brazil is the sole nation within the region where Bangladesh maintains an embassy, with its mission situated in the city of Dhaka. In general, Brazil is consistently positioned within the top ten countries in terms of its export volume to Bangladesh. However, Bangladesh faces challenges in meeting the demand for exports due to insufficient quantities. Moreover, Bangladeshi products encounter competition from Chinese and Indian products within the Brazilian market. In the present context, the adoption of effective economic diplomacy and the cultivation of robust bilateral relationships are imperative for the purpose of augmenting Brazil's export activities. To establish the foundation for future trade deals.

MERCOSUR, the most prominent economic market in Latin America, is a regional integration initiative. With that approaching method, Bangladesh will encroach on the market with its exportable items, including ready-made garments (RMG), pharmaceuticals, ceramics, leather, melamine products, and IT products. Bangladesh experiences upbeat about achieving its objectives, given its robust bilateral links with Argentina and Brazil, two influential and founding members of the Latin American market. Furthermore, Bangladesh is actively working towards enhancing these relations.

In the context of Latin America, the United States is exerting a significant influence through its implementation of the nationalistic policy known as 'America First.' Concurrently, China is establishing itself as a 'benign power' by actively engaging in Free Trade Agreements (FTAs) and Foreign Direct Investment (FDI). This dynamic between the two major powers can be understood as a manifestation of geopolitical rivalry. It is evident that within the context of this rivalry, there exists a potential for Bangladesh to capitalize on opportunities arising from both parties, namely in the realms of commerce and business. China might want Bangladesh to join its bloc, and the United States can do the same in the region after fostering bilateral ties with LAC. Both sides are expected to engage in amicable conduct towards Bangladesh. Consequently, Bangladesh will be granted greater commercial privileges in Latin America compared to countries such as Japan, India, and the European Union.

In the age of globalization, economic diplomacy has become a pivotal tool for states to deepen and expand their bilateral relations with other countries. This involves not only the promotion of trade and investment but also, and perhaps more importantly, the fostering of mutual trust and understanding. As we move deeper into the twenty-first century, the importance of forging robust economic ties cannot be understated, especially for developing nations looking to tap into newer markets and diversify their economic portfolios. The Republic of Bangladesh, being one of the fastest-growing economies in Asia, is no exception to this trend. However, while Bangladesh's economic diplomacy efforts with its immediate neighbors and major global players are well-documented, there is a relatively untouched dimension worth exploring: its economic relationship with the countries of Latin America.

Historically, Bangladesh and Latin American countries have had limited political and economic engagements. Despite the vast geographical distance and apparent cultural differences, there exists untapped potential for mutual cooperation. The question then arises: how can economic diplomacy serve as a bridge to deepen bilateral relations between Bangladesh and Latin American nations? As the world's political and economic landscapes continue to shift, it's more important than ever to grasp the nuances of various bilateral relationships. This research attempts to add to the ongoing conversation on international relations by studying the function of economic diplomacy between Bangladesh and Latin American countries.

1.2 The Phenomenal Growth in Maritime Technology: Pre-discussion on Inter-continental Trade

The natural and permanent proximity between Latin America and Bangladesh is a question of geography. However, in today's age of globalization and technology, merchants no longer have to worry about bridging such a large distance. In particular, the number of interactions between Brazilians and Argentines and Bangladeshis of all walks of life is growing steadily. The vast majority of social media platforms are instrumental in this. Multiple Bangladeshi news outlets report that young people in the country are getting married to their counterparts in countries like Peru, Mexico, Brazil, and Argentina which will assist Latinos' relations with Bangladesh to strengthen over time. The Internet has made the entire world more accessible than ever before. As a result, you won't have to spend any extra time educating yourself on the political, social, and cultural climates of each new destination.

The vast improvement in technology has been most noticeable in port communication systems. The elimination of key barriers to international trade is a direct result of advancements in maritime communication. The sea is the most dependable mode of transport for international trade. Exploring the globe's oceans and territories has been a respected occupation since humankind's earliest days. The organization's primary goal was to fortify itself through the exploitation of mineral resources. Now that navigation technology and accessories have advanced, contemporary seaports can facilitate trade across oceans.

Maritime transportation significantly influences the global economy, yet it does not receive prominent social and public acknowledgment, despite its numerous advantages, particularly in relation to the everyday lives of individuals, as a substantial portion of their possessions have been transported by sea. The industrial and economic progress of nations is largely facilitated by international trade, a significant portion of which relies heavily on sea transportation. In recent years, a significant volume of cargo, amounting to millions of tones, has been moved globally along this particular route. This transportation activity has made a substantial contribution, amounting to billions of dollars, to the overall global economy. A significant proportion of global trade is conducted by maritime transportation. The transportation of large volume and bulk cargoes is commonly facilitated using this method due to its cost-effectiveness, which is contingent upon factors such as extensive distances, substantial quantities, and significant weights of the goods involved. Human beings have developed several modes of transportation in order to surmount the obstacles and geographical distances that divide them from other nations and continents. The individual has utilized sea travel as a versatile instrument to pursue conquest, build connections, reinforce, and advance economic objectives via both domestic and international trade. Throughout the course of human history, individuals have consistently exhibited the inherent necessity to engage in the act of relocation, thereby transitioning from one geographical location to another. Transportation has played a crucial role in facilitating human endeavors such as exploration, colonization, conquering, and the interchange of cultural and commercial goods. There exist geographical regions that were initially geographically isolated due to topographical features but were subsequently connected by the formation of a body of water. The origins of maritime transportation may be traced back to approximately 3500 BC. During that particular period, limited cargo quantities and a small number of seamen were able to be transported along brief coastal routes. As technology progressed and new materials were introduced, the distances between objects increased significantly.

The commencement of the sixteenth century marked the initiation of a significant epoch in the realm of global maritime transportation, characterized by the momentous expeditions of exploration. The exploration of alternative maritime routes to remote nations necessitated the utilization of swift vessels, such as caravels, capable of traversing long distances across the open sea. Due to their inherently hazardous nature and the vast distances involved, these expeditions included venturing into uncharted territories of considerable uncertainty.

The Evolution of Maritime Trade: From Ancient Navigation to Modern Port Community Systems

Both technological developments and shifting economic priorities have contributed to maritime navigation's dramatic development over the centuries. The compass was the standard navigating tool for centuries, and it was widely thought that Arab traders from China brought it to Europe in the twelfth century. In the fifteenth century, Jacob's staff became commonplace, and it was used to measure the angle between the horizon and the stars; however, the sextant of the eighteenth century was more accurate. Together, these instruments and nautical tables allowed for precise determinations of latitude. However, John Harrison's chronometer (1759), a precision watch that allowed for accurate longitude calculations, radically altered the course of navigation.

This navigational prowess, coupled with the introduction of motorized ships, led to an explosive growth in sea-based trade over two centuries. As vessels became more efficient, ports emerged as vital hubs, knitting together a vast global network of trade. The last half-century has witnessed a quantitative surge in maritime trade, underpinned by increasing specialization. Vessels, ranging from bulk carriers, container ships, tankers, to cruise ships, have diversified, each tailored for specific cargo categories. Concomitantly, ports have metamorphosed from mere "gateways" to pivotal nexuses in global logistic chains, necessitating standardization and efficiency (Smith 2010). Among the plethora of vessel types, several warrant special mention. Bulk carriers transport substantial solid cargoes, leveraging cranes for load manipulation. The innovation of the container ship in the 1950s, attributed to Malcolm Mclean, transformed maritime transport, emphasizing safety, profitability, and environmental sustainability. Tankers, specializing in liquids and gases, transport a medley of chemicals, oils, and their derivatives. Refrigerated ships cater to perishable goods, ensuring optimal temperature control. Meanwhile, rolling load vessels are ingeniously designed to transport wheeled merchandise, equipped with systems to ensure stability. Lastly, cruise ships, floating behemoths, serve the burgeoning tourism sector, offering leisurely voyages across multiple ports.

Transitioning to modern times, the late 1970s heralded the advent of the Port Community System (PCS) in Bremen, replacing cumbersome paper documents with electronic counterparts. Over subsequent decades, major European ports embraced PCS, recognizing its potential to streamline logistics. Today, the indispensability of PCS is globally acknowledged, with many ports integrating IT infrastructures to enhance operational efficiency (Davies 2018). Such systems consolidate information from the myriad stakeholders involved in maritime trade, drastically reducing bilateral contacts and information disparities. Port Community platforms, by fostering data exchange and visibility, have become the bedrock of contemporary port logistics, offering unparalleled efficiency and value to port operations. The advancements in the global communication system have had a profound impact on the trajectory of international trade. Hence, the influence of distance on contemporary international trade is negligible. The significance of international maritime law and security systems is becoming increasingly vital in bolstering global economic collaboration. The geographical separation between Latin America and Bangladesh doesn't pose a significant obstacle to trade and technical advancement. Both India and China have made substantial advances in this particular domain. Hence, the Latin American market is poised to perceive Bangladesh as a significant and promising prospect. The essential factor for achieving the desired outcome is the genuine intention of the individuals responsible for formulating policies. If Bangladesh is able to effectively utilize this opportunity, it is likely to have a favorable outcome for the domestic economy of the nation. Furthermore, Bangladesh is determined to achieve its objective of attaining developed economic status.

1.3 Expectations for the Study's Exploration

This study seeks to shed light on an under-researched area of international relations. As both Bangladesh and Latin American countries seek to diversify their economic partners and look for sustainable growth opportunities, understanding the role of economic diplomacy in this dynamic becomes imperative. The outcomes of this study could provide policymakers, diplomats, and business leaders with insights and strategies to enhance bilateral engagements. Enhancing economic diplomacy between Bangladesh and the Latin American region holds substantial promise for both sides. This research proposal aims to provide policymakers, diplomats, business leaders, and stakeholders with a deeper understanding of the prospects of economic diplomacy and offer actionable strategies to strengthen bilateral economic relations. As we all know that Bangladesh wants to become a developed economy country by 2041. In such circumstances, the country cannot remain economically isolated from a continental region of the globe.

Bangladesh's neighboring country India has already vehemently economic connections with 13 Latin American countries. Moreover, the countries of this region can be economic partners of Bangladesh to reduce the dependence on global powers. Among these countries, Brazil aspires to be one of the most powerful countries in the world. So, it is important to establish international relations with Bangladesh. The geographical and cultural distance can be bridged by technological excellence through far-reaching economic diplomacy. This will create an opportunity to develop a direct relationship between Bangladeshis with the people of that region, which has already started the connection through the support of Bangladesh in the football game.

By capitalizing on untapped opportunities and addressing existing challenges, Bangladesh can expand its trade and investment ties, diversify its economic partnerships, and secure access to new markets. Similarly, Latin American countries can benefit from Bangladesh's expertise in sectors such as textiles, pharmaceuticals, information technology, and agricultural products, fostering economic growth and development. Strengthening economic diplomacy between Bangladesh and the Latin American region can contribute to mutual prosperity, increased regional integration, and the achievement of sustainable development goals.

The economic potential of the Latin region along with Bangladesh will be highlighted in this research thesis. In order to implement these possibilities, various aspects of a possible provision of economic policy will be examined. Despite the vast potential of trade with this region, it has not developed due to various reasons in the last 50 years. Among these are geographical distance and cultural differences. An attempt will be made to review their rationale through research. Besides, in the era of super technology-based globalization, information-based discussions on various ways to remove trade barriers are one of the main research objectives. At the same time, this research documentation will analyze various policies to formulate a realistic outline of how Bangladesh will enrich trade with the region in the changing situation. All data will be reviewed to verify the usefulness of potential economic diplomacy.

Analyzing and enhancing trade relations between Bangladesh and Latin American countries hold significant potential for economic growth, job creation, and improved living standards in both regions. This research proposal aims to provide policymakers, trade organizations, and stakeholders with a comprehensive understanding of the current state of trade and the necessary strategies to overcome existing challenges. By identifying opportunities and implementing trade facilitation measures, both Bangladesh and Latin American countries can diversify their trade portfolios, enhance export competitiveness, and foster economic cooperation. Ultimately, a stronger trade relationship will contribute to sustainable development, poverty reduction, and shared prosperity for both regions.

The research will encompass the time frame starting from Bangladesh's attainment of independence in 1971 until the present, with the aim of examining the progression of its economic interactions with nations in Latin America.

Even while the research tries to be as thorough and objective as possible, it must be aware of its limitations. Recognizing these shortcomings gives the study more legitimacy and opens up possibilities for future investigation, which may fill up any gaps or cover topics that were not able to be thoroughly explored. Knowing the limits of any academic project is just as important as knowing its scope and purpose. This study aims to be upfront about its goals and any obstacles by outlining its increased scope and potential restrictions. This will help to establish the study's credibility and provide a clear road forward for future research endeavors.

1.4 Literature Review: Comprehending the Current Research and Scholarship on Economic Diplomacy in Latin American Nations

Economic diplomacy has gradually emerged as a fundamental tool in shaping international trade relationships. In the context of Bangladesh and Latin American countries, its role becomes even less pronounced given the largely untapped economic potential between these geographies. This literature review offers a synthesis of available academic work to shed light on the role of economic diplomacy in deepening these bilateral trade relations. Bayne and Woolcock (2016) identified economic diplomacy as a blend of traditional statecraft and modern trade and investment practices. They argued that nations employ economic diplomacy to not only advance their trade and economic interests but also to forge strategic partnerships and foster geopolitical alliances. This conceptualization aligns well with Bangladesh's aspirations to cultivate deeper trade ties with Latin America. Understanding the Bangladesh perspective, Kabir et al. (2018) highlighted the nation's drive to diversify its trade partners as a part of its 'Look East' policy. While much attention has been concentrated on ASEAN and East Asian nations, Kabir and Akter emphasized the latent potential of

Latin America, a region with which Bangladesh shares several economic synergies, notably in textiles, agriculture, and technology.

Diving into the Latin American perspective, Ross (2015) presented an extensive analysis of Latin America's shifting trade dynamics post the commodities boom. Their research underscored the need for Latin American nations to diversify trade partners and engage with rapidly growing economies, like Bangladesh. He particularly emphasized the opportunities in sectors such as agro-industry, textiles, and IT services. However, the potential isn't without challenges. Porimol Palma (The Daily Star 2023); Towkir Hossain (The Business Standard 2023) dissected the logistical, cultural, and regulatory barriers impeding robust trade ties between Bangladesh and Latin American nations. He pointed to the absence of direct shipping routes, significant cultural and linguistic differences, and diverse regulatory landscapes as significant impediments. Yet, they concluded optimistically, advocating for economic diplomacy as the linchpin that could address these challenges head-on.

Supporting this view, Bdnews24 (2005) presented a news report on the success of economic diplomacy between Bangladesh and Chile. They demonstrated how targeted trade dialogues, investment forums, and cultural exchange programs, orchestrated under the umbrella of economic diplomacy, led to a 20% increase in bilateral trade within just two years. Their investigation provides a compelling blueprint for replicating similar successes with other Latin American nations. Finally, Wafiur Rahman (2021, Dhaka Tribune) argued that the future of Bangladesh-Latin America relations hinges on multitiered economic diplomacy. He proposed a blend of government-led initiatives, business-to-business engagements, and people-to-people interactions. By leveraging this triad, Wafiur Rahman believes that both Bangladesh and Latin American countries can overcome extant barriers and realize the full spectrum of their trade potential.

The available literature compellingly highlights the crucial role of economic diplomacy in nurturing and deepening trade relations between Bangladesh and Latin American countries. While the potential is vast, nuanced and multipronged diplomatic endeavors are essential to navigate the challenges and forge a sustainable economic partnership. Economic diplomacy has emerged as a critical strategy for nations in the contemporary globalized world. Specifically, the relationship between Bangladesh and Latin American countries, though under-researched, presents a plethora of opportunities that economic diplomacy could unlock. This literature review synthesizes key scholarly works to explore this intricate interplay.

Rahman and Siddiqui (2018) provided a comprehensive analysis of Bangladesh's Ready-Made Garment (RMG) sector and its potential in the Latin American market. They emphasized the importance of economic diplomacy in navigating regulatory and logistical challenges, suggesting that dedicated trade missions could bridge information gaps. Towkir Hossain (2023) investigated the significance of multilateral forums. As a means of fostering mutual understanding and cooperation between Bangladesh and Latin American states, they advocated the use of forums like the United Nations Economic Commission for Latin America and the Caribbean (ECLAC) as instruments of economic diplomacy. Rafikul Islam (2023) report at The Business Post on the Bangladeshi jute industry's prospects in Latin America provided a sector-specific insight. He advocated for targeted economic diplomacy initiatives, such as B2B meetings and participation in specialized trade fairs, to promote this eco-friendly product in environmentally conscious Latin American markets.

The extant literature elucidates the transformative potential of economic diplomacy in deepening trade relations between Bangladesh and Latin American countries. While challenges remain, the opportunities are vast, underscoring the need for sustained and strategic efforts. Economic diplomacy has increasingly been recognized as an essential instrument in the toolbox of nation-states, especially in the evolving global economic landscape. This literature review synthesizes key works to understand the interplay between economic diplomacy and bilateral relations, emphasizing its multifaceted impact and implications. van Bergeijk (2009) elaborates further on the dimensions of economic diplomacy, pointing out its two primary avenues: commercial diplomacy, which pertains to trade and investments, and financial diplomacy, focusing on monetary issues and financial flows. This distinction is critical; as different diplomatic strategies may be deployed based on the specific economic interaction at play.

The role of economic diplomacy in forging bilateral relations isn't merely a reflection of existing economic ties but can actively mold and influence the trajectory of bilateral relationships. Research on European-Latin American relations emphasizes how multinational corporations, industry associations, and even cities play vital roles in shaping economic relations, often bypassing traditional state-centric channels. A more recent study by Benchea (2023) delves into the digitalization of economic diplomacy. They contend that in the age of information, big data analytics, social media platforms, and digital networks are transforming how states practice economic diplomacy, offering both new tools and challenges.

In the realm of scholarly inquiry, a number of studies have been conducted pertaining to the economic diplomacy of Bangladesh. In the work published in 2003, Absar provided a comprehensive analysis of Bangladesh's economic diplomacy within the context of the World Trade Organization (WTO) era.

The correlation between economic diplomacy and the foreign policy of Bangladesh, highlights the significance of Southeast and East Asia in the country's economic and commercial diplomacy. It is imperative for the foreign policy of Bangladesh to prioritize economic diplomacy in order to leverage geo-strategic and geo-economic opportunities arising from the upcoming era of multilateralism. Facilitating inward foreign direct investment (FDI), accessing contemporary technical and technological expertise, enhancing commerce, and receiving inward official development assistance (ODA) would be advantageous for Bangladesh.

The provided literature review offers limited understanding of economic diplomacy's role in shaping bilateral trade relations between Bangladesh and Latin American countries. Yet, like all academic works, there exist research gaps that future studies can address.

Sector-Specific Analysis: While general trade dynamics have been discussed, there seems to be a lack of sector-specific analysis. Future research could delve deeper into particular industries, for example, how the Bangladeshi textile industry can collaborate with specific Latin American countries, or how Latin American agro-products might find a market in Bangladesh.

Comparative Studies: While there's a case study on Bangladesh-Brazil economic relations, more comparative studies with other countries in the region can offer diverse perspectives and a richer understanding of bilateral potentials.

Micro-Level Barriers: In the discussion of logistical, cultural, and regulatory barriers, there might be micro-level challenges at the firm or industry level that haven't been explored. These might include issues related to business practices, supply chain intricacies, or consumer preferences.

Role of Non-State Actors: The literature predominantly emphasizes state-led initiatives. The potential role of non-state actors, such as multinational corporations, NGOs, and civil society in economic diplomacy, remains largely untouched.

Long-Term Impact Analysis: Bdnews24 reported a 20% increase in trade in two years with Chile. But what about the long-term sustainability of such growth? Future research could study the long-term impacts and sustainability of such economic diplomacy initiatives.

Digital Diplomacy in Trade: In an era of rapid digitalization, how might digital diplomacy tools be used to foster trade relations? This is an emerging field, and its specific relevance to Bangladesh-Latin American relations remains unexplored.

Economic Diplomacy Training and Skill Development: How well-equipped are diplomats and trade negotiators from both sides to navigate this evolving landscape? There's a gap in the literature examining the training and skills aspect of those involved in economic diplomacy.

Impact of Global Dynamics: How might global events, like the US-China trade tensions, Brexit, or global recessions, influence Bangladesh-Latin American trade ties, and how can economic diplomacy navigate such challenges?

Stakeholder Perceptions: A qualitative study capturing the perceptions, experiences, and recommendations of stakeholders directly involved in trade (e.g., exporters, importers, trade facilitators) between Bangladesh and Latin American countries could provide valuable insights.

Policy Recommendations and Implementation: While challenges and opportunities are identified, there seems to be a gap in providing specific policy recommendations and strategies for their implementation.

Environmental Considerations in Trade: As global emphasis on sustainable practices grows, there's a lack of literature examining how environmental considerations might influence economic diplomacy and trade negotiations between Bangladesh and Latin America.

Socio-Cultural Exchanges and Their Economic Impacts: While the literature touches upon cultural barriers, there's scant exploration on how socio-cultural exchanges could potentially boost trade relations by fostering trust and understanding.

Impact of Regional Trade Agreements: There's a gap in understanding how regional trade agreements, such as Mercosur in Latin America, might influence bilateral trade with countries outside the bloc, like Bangladesh.

Technology Transfers and Knowledge Sharing: With both regions having their unique technological strengths, the potential for technology transfers, collaborative R&D, and its subsequent impact on bilateral trade remains under-researched.

Role of Diaspora: Considering the diaspora's increasing influence in global trade dynamics, how might the Bangladeshi diaspora in Latin America (and vice versa) act as catalysts for economic diplomacy and trade?

Investment Diplomacy: While trade is often the focus, foreign direct investment (FDI) plays a crucial role in economic relations. Literature specifically examining the investment climate, potential sectors, and challenges for FDI between Bangladesh and Latin American nations is sparse.

Evaluation of Existing Bilateral Agreements: If there are existing frameworks or agreements, a systematic evaluation of their efficacy, challenges faced, and areas of improvement can be a rich area of study.

Addressing these gaps with detailed research can provide invaluable insights for policymakers, businesses, and other stakeholders, enhancing the efficacy of economic diplomacy between Bangladesh and Latin American countries. Identifying these gaps can guide future research endeavors to ensure a holistic understanding and more effective promotion of trade relations between Bangladesh and Latin American countries.

1.5 Approaches

The nature of analytical research was combined with a mixed-method qualitative resource to inform these policy talks. The perspectives of concerned individuals from various categories, such as experts in policy, professors, government officials, diplomats, and professionals, are gathered through online and offline interviews. In addition to these components, policies and opportunities are examined through the use of secondary data analysis and observation methods. In the meanwhile, the study's goal is to conduct exploratory, applied research.

1.6 Conclusion

Foreign policy, intrinsically intertwined with domestic policy, fundamentally represents a sovereign state's strategy to achieve its objectives in interactions with other nations. Bangladesh, a country that gained its independence in a remarkable episode of global politics in 1971, sets its foreign policy tenets based on its constitutional provisions. These dictate the promotion of equality, respect for national sovereignty, non-interference in other countries' internal affairs, peaceful resolution of disputes, adherence to international law, and commitment to the principles of the UN Charter. Celebrating 52 years of statehood, Bangladesh has established diplomatic relations with nearly 60 out of 198 recognized countries, leaving a significant 65% of nations unengaged. Crucially, regions such as Latin America and Africa remain largely uncharted territories in Bangladesh's diplomatic endeavors. Economic diplomacy, a pivotal mechanism in sculpting international economic relationships, has witnessed Bangladesh's burgeoning interest as the nation seeks to expand its global economic presence.

However, when we zone in on Latin America, Bangladesh's engagement is strikingly minimal. Only three countries—Brazil, Mexico, and Argentina—have seen any substantive international exchange. Historical events, such as the cessation of partnership with Argentina in 1978 due to economic constraints, have left imprints on these relationships. Yet, the reopening of Argentina's embassy in Bangladesh offers a glimmer of hope and signifies the latent potential of trade growth and diversification between the two regions. Key areas ripe for bilateral cooperation include grain, edible vegetable oils, automotive industries, pharmaceuticals, agrochemicals, garments, IT, skilled manpower, arts, culture, research, and higher studies. Understanding Latin America requires acknowledging its shared history of conquest, colonization, and subsequent independence movements. Comprising the entirety of South America, Mexico, Central America, and certain Caribbean islands, Latin America is home to 13 independent nations and approximately 660 million people. Major economies like Brazil and Mexico dominate the region. Bangladesh could draw inspiration from the likes of India and China, which have augmented their trade partnerships with Latin America by nearly 25%. Such a partnership, given the region's vast consumer market, would be strategic for Bangladesh in solidifying its economic stature while paralleling its neighbors. A review of the trade trajectory between Bangladesh and Latin America indicates promising growth over the past decade. By 2020, bilateral trade amounted to around 3 billion USD, highlighting the increasing economic rapport between the two regions. While Bangladesh exports garments, textiles, pharmaceuticals, ceramics, and jute products, it imports mainly mineral fuels, food items, chemicals, and machinery. But this is just scratching the surface of what could be a flourishing trade relationship. Challenges like high tariffs, non-tariff barriers, logistical issues, limited market insights, and cultural nuances are barriers that need addressing.

While Bangladesh has made strides in establishing itself on the global stage, there's an evident untapped potential in its relations with Latin America. Addressing existing challenges, re-evaluating historical ties, and formulating an informed and proactive economic diplomacy can pave the way for a robust and mutually beneficial relationship between Bangladesh and the nations of Latin America. This relationship could significantly influence the future trajectory of Bangladesh's economic and diplomatic landscapes.

To accomplish overall foreign policy goals, states often adopt a multifaceted strategy known as 'economic diplomacy,' which centers on encouraging trade, investment, and economic cooperation. As the scope of globalization has broadened over the past few decades, the role of economic diplomacy in shaping the development of bilateral ties has grown in importance. The necessity of trade and investment diversification and strong alliances is becoming increasingly apparent to countries around the world. Bangladesh is in a prime position to take advantage of these developments as its economy continues to expand fast. A whole hemisphere of opportunity, Latin America, remains untapped despite its persistent pursuit of economic diplomacy with its neighbors and several key world powers. Economic diplomacy's impact on bilateral relations between Bangladesh and Latin American countries is ripe for investigation, and it's likely that such an investigation will yield complex insights. The significance of such varied connections will definitely increase if the global order becomes more multipolar. Bangladesh and Latin American countries could benefit economically by working together, despite their great distance from one another and apparent cultural and historical differences. With so many untapped resources and opportunities for cooperation, both regions have much to offer each other.

The ever-evolving global economic landscape underscores the importance of nations diversifying their trade and investment partnerships. Bangladesh, a rising economic power in South Asia, has predominantly focused its international economic engagements within its immediate vicinity and with established global powers. However, in order to achieve its economic goals, the country must boldly go where it has never gone before in terms of working with other countries; one such region is Latin America. Opportunities for commerce, investment, and technology cooperation abound in Latin America, which is home to economies as varied as Brazil's and the growing markets of Chile and others. Since Bangladesh and Latin America share relatively little in the way of shared political and historical baggage, they have a rare opportunity to forge new partnerships based on the shared economic benefits they provide each other.

However, there has been surprisingly little study of the economic diplomacy ties between Bangladesh and Latin American nations. Given the mutual benefits that could accrue to both regions, it is remarkable that there is a lack of discussion about them in academic and policy circles. Therefore, the purpose of this research is to shed light on the possibilities, obstacles, and intricacy of economic diplomacy between Bangladesh and Latin America. By doing so, it aspires to serve as a guide for politicians, academics, and business leaders who are keen on fostering ties between these two thriving areas.

References

Absar SS (2003) Economic diplomacy for Bangladesh: a prescription for less developed countries in the World Trade Organisation era. South Asia J South Asian Stud 26(3):349–357. https://doi.org/10.1080/0085640032000178925

Bayne N, Woolcock S (Eds.) (2016) The New economic diplomacy: decision-making and negotiation in international economic relations (4th ed.). Routledge. https://doi.org/10.4324/9781315555188

Benchea L (2023) Impact of digitalization on economic diplomacy: mega-trends and opportunities. Rom Econ J, 85. https://doi.org/10.24818/rej/2023/85/02

Davies R (2018) Port community systems: transforming maritime operations. Nautical Stud J

Hossain T (2023) The shifting trends in Bangladesh and Latin America relations. [online] The Business Standard. Available at: https://www.tbsnews.net/thoughts/shifting-trends-bangladesh-and-latin-america-relations-591298. Accessed 5 Oct 2023

Islam R (2023) Dhaka looks to tap export potential. [online] businesspostbd.com. Available at: https://businesspostbd.com/front/dhaka-looks-to-tap-export-potential-2023-01-14. Accessed 7 Oct 2023

Kabir R, Akter S, Islam R, Molla M (2018) Justification of look east policy and Bangladesh. Int J Adv Res 6 (Dec). 673–683 (ISSN 2320-5407). https://www.journalijar.com

Rahman MH, Siddiqui SA (2015) Female RMG worker: economic contribution in Bangladesh. Int J Sci Res Publ 5(9) (ISSN: 2250-3153). http://www.ijsrp.org/research-paper-0915.php?rp=P454564

Ross M (2015) Conflict and natural resources: is the latin American and caribbean region different from the rest of the World?

Smith J (2010) Evolution of maritime trade routes: from sails to containers. Oxford Maritime Press

van Bergeijk (2009) Economic diplomacy and the geography of international trade. Edward Elgar Publishing Limited, Cheltenham

Chapter 2
Economic Diplomacy and Bilateral Relations: Conceptual Clarity

Abstract Economic diplomacy has evolved as a pivotal instrument in advancing bilateral relations, intertwining economic interests with foreign policy objectives. The theoretical foundations of this practice, rooted in realism, liberalism, and constructivism, provide a framework for analyzing its implications in reshaping Bangladesh's diplomatic and economic engagement with Latin American nations. This study delves into the conceptual clarity and theoretical aspects of economic diplomacy to explore its potential in fostering reciprocal growth and strategic partnerships between Bangladesh and Latin America. Realism underscores economic diplomacy as a strategic tool for achieving national interests, emphasizing power dynamics and relative gains. Bangladesh's pursuit of stronger ties with Latin America reflects this perspective, as it seeks to enhance economic security by diversifying trade routes and reducing dependence on traditional markets. Conversely, liberalism highlights mutual benefits and interdependence, advocating for collaborative frameworks like bilateral trade agreements, investment treaties, and technology transfers to foster shared prosperity. Constructivist theories emphasize the role of ideas, norms, and shared values in shaping economic diplomacy. Bangladesh's aspirations to build a knowledge-based economy and Latin America's focus on industrial and agricultural advancements align with this perspective, creating opportunities for cultural exchanges and sustainable collaborations. Institutions such as the WTO and MERCOSUR provide platforms for harmonizing these efforts, reinforcing the role of international organizations as enablers of economic diplomacy. This research also examines the multidimensional tools of economic diplomacy, including trade promotion, foreign direct investment (FDI) attraction, and the negotiation of economic treaties. For Bangladesh, integrating these tools into its foreign policy framework can address challenges such as logistical barriers, tariff discrepancies, and a limited diplomatic presence in Latin America. Similarly, leveraging soft power through cultural diplomacy and public engagement can bridge gaps in understanding, fostering deeper bilateral connections. The theoretical exploration further highlights the dynamic nature of economic diplomacy, with non-state actors such as multinational corporations and civil society playing increasingly influential roles. These actors can complement state-led initiatives by fostering business-to-business linkages and facilitating innovation-driven partnerships. Moreover, the liberal notion

M. T. Islam and M. B. Hossen, *Economic Diplomacy*, The Latin American Studies Book Series, https://doi.org/10.1007/978-3-031-81362-7_2

of complex interdependence suggests that economic ties between Bangladesh and Latin America can transcend traditional trade models, encompassing shared goals like climate resilience and digital transformation. By grounding its analysis in theoretical perspectives, this study underscores the strategic potential of economic diplomacy in redefining Bangladesh-Latin America relations. It concludes that a nuanced and adaptive approach, informed by historical patterns and contemporary global trends, can transform bilateral engagements into a cornerstone of mutual economic development and geopolitical stability.

Keywords International trade · Foreign policy · Economic relations · FDI · Bilateral relations

2.1 Economic Diplomacy

Economic diplomacy refers to the strategic use of foreign policy by nations in order to cultivate domestic economic power and promote prosperity. The concept of economic diplomacy encompasses diplomatic endeavors aimed at advancing a nation's economic interests. Additionally, it encompasses the practice of employing economic resources in diplomatic endeavors with the aim of accomplishing a particular objective in foreign policy. Economic diplomacy, sometimes termed commercial diplomacy, is a vital tool for nations seeking to bolster their economy in today's globalized environment. It primarily emphasizes export promotion and attracting foreign direct investments. At its core, economic diplomacy navigates international economic challenges to foster national prosperity. This is achieved by leveraging government resources to amplify trade, facilitate investments, and engage in bilateral and multilateral trade negotiations.

It also encompasses strategies like foreign aid and economic sanctions to further a country's foreign policy. The modern landscape of economic diplomacy highlights collaboration between official state agencies and non-governmental entities. As economic diplomacy evolves, private sectors and civil societies increasingly influence a country's negotiation stance. While traditional diplomacy operates on bargaining power, today's economic diplomacy prioritizes negotiation skills and fostering interdependence between states, leading to peaceful relations. A shift in focus is also evident, with an emphasis on acquiring knowledge and skills rather than just technology or projects. This approach not only aids in economic development but also nurtures indigenous industries, technology, and skills. In essence, economic diplomacy is about fostering friendships through trade and investment.

2.2 Economic Diplomacy as a Modern Concept: Background and Significance

Economic diplomacy revolves around the strategies and activities nations employ to drive their cross-border economic interests. These activities are facilitated by both state and non-state actors and primarily focus on promoting foreign business, investments, technology, and tourism. Economic diplomacy at the state level can be understood in three dimensions:

Bilateral Economic Diplomacy: Using political influence to promote bilateral trade and investment, streamline market functions, and reduce transaction costs. This includes export promotion and state visits.

Economic Security: Leveraging economic assets to enhance the benefits of cooperation and maintain stable political relationships. This can involve structural policies, trade agreements, and sometimes political trade distortions like embargoes.

Trade Diplomacy: Establishing a conducive political and economic environment on an international scale, largely facilitated by organizations like the WTO, OECD, and the EU.

Though sometimes used interchangeably, "economic diplomacy" and "commercial diplomacy" serve different aims but have some characteristics. Sridharan considers commercial diplomacy economic diplomacy. Commercial diplomacy focuses on inter-state trade and transactions, but these interactions are usually based on bilateral treaties between the two States. Advocates of the term 'commercial diplomacy' include elements of 'economic diplomacy' in it. Ruel says the debate over these two terms may never end:

> … commercial diplomacy is the international relations of business that knits together political and entrepreneurial activities and agents in the global market.

Lee and Ruel say commercial diplomacy promotes trade and investment by building diplomat-commercial networks. It is believed that economic diplomacy may achieve economic and foreign policy goals and commercial diplomacy is essential to economic diplomacy. So, economic diplomacy is broader than commercial diplomacy. Economic diplomacy as a way to build relationships and maximize national gains in trade, investment, and other economic exchanges. However, economic diplomacy can take over from commercial diplomacy. Commercial and economic diplomacy aims for significant "reciprocity" between chosen States based on bilateral accords, despite controversy. Many international relations and economics professors have studied economic diplomacy. Academic definitions of economic diplomacy with references:

Nicholas Bayne (2008): 'Economic diplomacy deals with the nexus between power and wealth in international affairs. Economic diplomacy not only promotes the state's prosperity but also, as occasion demands and opportunity permits, manipulates its foreign economic and financial relations in support of its foreign policy—as such, it is a principal tool of statecraft.'

Rana (2007): 'Economic diplomacy refers to the decision-making, policymaking, and advocating activities of diplomats in charge of bilateral, regional or multilateral negotiations, agreements, treaties and conventions that address issues of mutual interest in order to reduce the risk and enhance economic and commercial relations.'

Saner and Yiu (2003): 'Economic diplomacy is the employment of international economic instruments to achieve the national objectives of a country. These instruments include, but are not limited to, official development aid, international trade, investment treaties, and other forms of economic and financial negotiations.'

Lee and Hudson (2004): 'Economic diplomacy is the art of negotiating and networking by governments to realize commercial gains for national companies, trade associations, or other organized economic interests in global markets.'

Since the beginning of the 2000s, economic diplomacy and security have become central to global policy discussions. This shift is due to factors like the integration of formerly communist nations into the global economy, unprecedented price hikes in essential goods like oil, the challenges posed by climate change, and the emergence of transnational terrorism. The present era is marked by deepening ties between nations and individuals worldwide, unlocking vast economic potential in emerging markets, piquing the interest of both researchers and policymakers in economic diplomacy.

Figure 2.1 showcasing the surge in international interactions in economic diplomacy during the latter half of the twentieth century highlights two key trends: the frequency of state visits and the use of economic sanctions like boycotts and embargoes. By 2000, economic sanctions aiming at foreign policy objectives were four times higher than in the 1950s. Similarly, state visits by Heads of State from the US, Germany, and France saw an eightfold rise during the same timeframe.

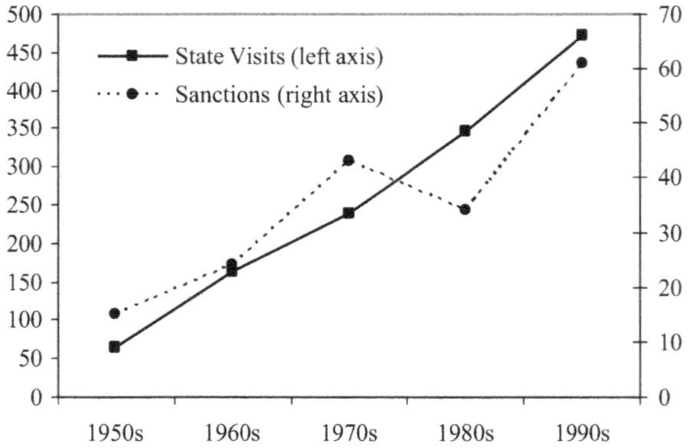

Fig. 2.1 Trends in economic diplomacy 1950–1999. *Source* Nitsch (2007, p. 1800, Table 1); all visits by Heads of State from France, Germany and the United States

2.3 Evolution of Economic Diplomacy

Economic diplomacy has its roots in the early exchange of goods, exemplified by ancient practices such as those seen in Ancient Greece. Its significance can be traced back to the emergence of consulates tied to trade, with these consulates serving as precursors to modern diplomatic missions. This paper reviews the evolution of economic diplomacy from ancient practices to its significance in the nineteenth and twentieth centuries.

2.3.1 Antecedents in Ancient Times

Ancient trade activities, such as those observed in the fairs of Ancient Greece, fostered diplomatic relations. The Delian League, which peaked in the early seventh century BCE, is a testament to this. The concept of resident missions, which became prominent by the late fifteenth century AD, had significant economic undertones, originating from European consulates established due to international trade needs.

2.3.2 Economic Diplomacy from 1500s

From the 1500s to World War I, diplomatic missions were largely dominated by high politics, although the significance of economic diplomacy persisted. Reasons include: The persistence of merchant-established consular posts. Exponential growth in international trade by the eighteenth century further intensified in the nineteenth century. The pivotal role of diplomatic missions in negotiating commercial treaties. The late nineteenth century saw a resurgence of economic diplomacy, propelled by European colonial ambitions in search of raw materials and markets.

2.3.3 Nineteenth and Twentieth Century Economic Diplomacy

Throughout the nineteenth and twentieth centuries, consular services underwent reformation and professionalization. Diplomatic missions began directly supporting business ventures abroad, with ambassadorial figures and their aides leading commercial treaty negotiations. Major capital investment projects, driven by broader political interests like national prestige or control over strategic resources, became priorities. The emergence of commercial attachés in the late nineteenth century signified a shift towards promoting daily trade activities. The UK and the US, among others, adopted various administrative models to manage these commercial staff.

2.3.4 Late Twentieth Century Economic Diplomacy

Events such as World War I, the economic recessions, and the rise of economic nationalism in the 1930s bolstered the emphasis on economic diplomacy. Post-WWII saw a relentless rise in economic priorities. The Cold War's end and the dissolution of the European empires brought forth 'developing countries' that prioritized economic goals. With advancements in transport and communication in the 1960s, foreign ministries faced pressure to justify their existence and budgets. To this end, they sought support from big businesses, SMEs, and trade associations, underscoring their commitment to economic diplomacy. Economic diplomacy's trajectory from ancient trade activities to its prominence in modern times demonstrates its enduring relevance. As history shows, economic interests have continually intertwined with diplomatic efforts, reflecting the intricate balance between commerce and international relations.

2.4 Patterns and Techniques of Economic Diplomacy

Economic diplomacy operates at three levels: national, regional, and international, each demanding distinct tactics and strategies. At the national level, it seeks mutually beneficial arrangements to enhance foreign policy relations between states, emphasizing reciprocity. However, reciprocity might not always be equal, especially when involving a developing nation. Economic diplomacy also emphasizes capacity-building to fortify a nation's economic foundation, minimizing economic dependence. Internationally, economic diplomacy focuses on establishing framework regulations, such as conventions or resolutions. Unlike bilateral efforts, international economic diplomacy is not driven by national interests but by broader global objectives. It necessitates a forward-looking vision and a comprehensive understanding of regulatory requirements.

Historically, post-decolonization, many nations continued economic ties with their former colonizers. While historical ties play a role in economic diplomacy, there's a growing emphasis on diversifying relations and seeking new, innovative partnerships. Economic diplomacy has evolved from unilateral actions, where developing nations sought help from developed ones, to mutual exchanges. But, there's a need for developing nations to reevaluate the practice of reciprocating with natural resources.

Effective knowledge-sharing and resource-building, seen especially between two developed nations, highlights the significance of a knowledge-based economy in economic diplomacy. However, economic diplomacy isn't solely for international relationships; it should bolster internal socio-economic development. Regionally, economic diplomacy has grown in prominence over the past five decades, with regional economic integrations offering both advantages and challenges. A vital consideration is ensuring global market access, avoiding trade barriers. The ultimate goal of economic diplomacy is achieving national self-sufficiency, competitiveness in world markets, and strengthening other forms of rapport with states. However, for economic diplomacy to be effective, nations must also ensure democratic governance, protective legislation, and a reliable judiciary to foster investor confidence.

2.4.1 Changing Patterns of Economic Diplomacy

The dynamics of international economic diplomacy have experienced substantial changes since the 1980s. These transformations have been influenced by various factors, such as the economic ambitions of emerging nations, the increased agency of developing countries through their ownership of natural resources, the market power of technologically advanced nations, the formation of regional economic integrations, and the growing recognition of economic rights among developing nations. Several significant events have occurred to signify this transition, such as the adoption of the UN General Assembly Resolution on Permanent Sovereignty over Natural Resources in 1962, the establishment of the Charter of Economic Rights and Duties of States in 1974, and the issuance of the Declarations on the New International Economic Order in the same year. Historically, international economic diplomacy was characterized by economic domination during colonization and bilateral agreements to ensure freedom of navigation and commerce. However, decolonization shifted the paradigm, as newly independent states began asserting their economic sovereignty and rights over resources.

This newfound assertiveness became especially evident post-1974, with states emphasizing guidelines for foreign private investments, aiming for balanced mutual interests in negotiations, and incorporating provisions for capacity-building. Emerging states have since redefined economic diplomacy. Their progress in technology and private foreign investment cannot be sidelined by developed states, as these emerging markets pose business challenges to the developed world. With modern technology, developing nations have access to negotiation tactics and knowledge used by traditional powers, underscoring the importance of effective negotiation techniques in economic diplomacy. It's essential for diplomats to master the art of trade and investment contract negotiations. State contracts, based on bilateral investment treaties, present unique challenges. While transnational corporations negotiate with public entities representing their sovereign states, the power dynamics are inherently skewed given the authoritative stature of states as law-makers. Historical prejudices may have occasionally hindered developing nations in negotiations, but the contemporary scenario underscores mutual dependencies. Developed nations, in facilitating capacity-building in the developing world, can help ensure global prosperity and achieve true globalization.

The evolving pattern of economic diplomacy also encompasses issues like the Syrian crisis, with its multifaceted challenges affecting global trade, leading to larger economic implications. Economic diplomacy, inherently dynamic, requires multidisciplinary expertise. A well-conceived economic diplomacy aims at national interest protection while fostering international cooperation, recognizing that trade and investments cement inter-state relationships. Examples like ASEAN underline this shift, emphasizing capacity-building and local skill development as pivotal in foreign investment contracts.

In summary, economic diplomacy has transitioned from historical domination to a platform for mutual respect and cooperation. It holds the potential to promote peace and international collaboration through trade and investment, urging the need for binding standards and stronger international platforms for inter-state economic relations.

2.5 Basic Functions of Economic Diplomacy

Economic diplomacy, as an intersection of international economics and diplomacy, has adopted a variety of patterns and techniques over time. While these vary depending on a country's economic strength, strategic goals, and geopolitical contexts, certain prominent patterns and techniques have been noted in academic literature and practice.

1. **Trade Promotion and Facilitation**

A primary function of economic diplomacy is the promotion of trade. Economic diplomats work to showcase their country's products and services, encourage exports, and streamline the import of vital goods. Also organizing trade missions, participating in international trade fairs, and fostering business-to-business links.

2. **Negotiation of Economic Treaties**

This refers to efforts to establish formal economic relations with other countries through treaties. Engaging in bilateral and multilateral negotiations to forge trade agreements, investment treaties, and double taxation treaties. Pre-negotiation research, stakeholder consultations, and iterative bargaining are standard practices.

3. **Economic Sanctions and Incentives**

Sanctions are punitive economic measures taken against countries to achieve diplomatic goals, while incentives are economic rewards offered for alignment or cooperation. Implementing trade embargoes, freezing assets, or offering preferential trade terms and development aid.

4. **Promotion of Foreign Direct Investment (FDI)**

Economic diplomats strive to create an appealing environment for foreign investors, portraying their nation as a lucrative investment destination. Conducting investor roadshows, ensuring protective legal frameworks, and offering investment incentives such as tax breaks (UNCTAD 2019).

5. **Economic Intelligence Gathering**

Understanding economic landscapes of other countries can inform domestic policy decisions and trade strategies. Collaborating with businesses, think tanks, and international organizations to gather data on market conditions, regulatory environments, and potential risks (Saner and Yiu 2003).

6. **Dispute Resolution**

Economic disputes between countries, whether related to trade, investments, or other issues, are often mediated through economic diplomacy. Engaging in diplomatic dialogues, leveraging international arbitration mechanisms, and seeking mediation through international bodies like the WTO (Busch and Reinhardt 2003).

7. **Public Diplomacy and Branding**

Linking public diplomacy with economic objectives can enhance a country's brand image and increase its attractiveness as a trade and investment destination. Cultural exchanges, tourism promotion campaigns, and educational scholarships.

8. **Multilateral Economic Diplomacy**

In an era of globalization, economic diplomacy extends beyond bilateral engagements to multilateral forums. Actively participating in international forums like G20, APEC, and BRICS, influencing global economic norms and policies.

While these features do serve as a guide for economic diplomacy, it is important to remember that the success of any given strategy depends on the specifics of the situation, the state of the global economy, and the objectives of the countries involved.

In order to accomplish goals on the local, regional, and global scales, economic diplomacy employs a wide range of strategies. Here is a quick rundown of some of the most important tools used in economic diplomacy:

1. **Bilateral and Multilateral Negotiations**: Diplomats address economic concerns, such as trade barriers and joint ventures, through dialogues with single or multiple nations.
2. **Trade Agreements**: These stipulate trade conditions between nations, such as tariff reductions and standards (Hoekman and Kostecky 2009).
3. **Economic Sanctions**: The imposition of penalties for political objectives with the intent of modifying the behavior of the sanctioned country.
4. **Promotion of National Interests**: Increasing exports by advocating for domestic businesses and industries abroad (Reuveny and Thompson 2001).
5. **Investment Treaties**: These treaties encourage Foreign Direct Investment by promoting and protecting investments between nations (Sornarajah 2010). Economic Summits and Forums: Participation in international gatherings to shape global economic policies (Kirton 2013).
6. **Cultural and Academic Exchanges**: These programs foster economic ties and mutual understanding (Cummings 2003).
7. **Lobbying and Advocacy**: Engaging stakeholders in targeted countries to promote economic interests.

8. **Capacity-Building Initiatives**: Assisting nations in strengthening their economic skills and infrastructure (Scott and Wilkinson 2011).
9. **Utilizing International Organizations**: Engaging entities like the WTO to mold global economic norms (Goldstein et al. 2007).
10. **Trade Missions**: Exploring economic opportunities in foreign nations (Morrison et al. 2003).
11. **Economic Intelligence Gathering**: Informing national strategies by analyzing global economic trends.
12. **Public Diplomacy and Communication**: Using media to promote economic interests (Melissen 2005).
13. **Conflict Resolution Mechanisms**: Addressing economic disputes between countries.
14. **Regulatory Diplomacy**: Collaborating on shared regulations for smoother economic cooperation (Drake and Nicolaïdis 1992).

These techniques, when effectively executed, bolster a nation's economic stance, ensuring growth and fostering harmonious relations with partners.

2.6 Foreign Policymaking with Economic Diplomacy

Foreign policymaking, when intertwined with economic diplomacy, hinges on leveraging a country's economic strengths and vulnerabilities to achieve its international objectives. Economic diplomacy offers tools like trade agreements, bilateral and multilateral negotiations, and sanctions, which play a critical role in shaping a nation's external interactions. As Hoekman and Kostecki (2009) articulate, trade agreements, for instance, not only regulate tariffs and standards but also become instruments of foreign policy that can tighten or relax relationships with other countries.

Moreover, economic sanctions, emerge as powerful tools in foreign policy to influence or penalize nations without resorting to military force. However, the effectiveness of such tools depends on accurate economic intelligence and a clear understanding of global economic trends. Additionally, for foreign policy's success, underpinned by economic diplomacy, requires active engagement in international forums to shape global economic policies. In essence, economic considerations are central to a holistic foreign policy approach, emphasizing the need for integrating economic tools and strategies into foreign policy deliberations.

2.6.1 Economic Diplomacy and Foreign Policy: A Holistic Approach

Economic diplomacy extends beyond establishing business relationships, acting as an integral tool to reinforce a nation's infrastructure and intellectual prowess. It emphasizes a knowledge-based economy as a catalyst for creativity. Contrary to limited perceptions, economic diplomacy is equally essential between countries of analogous economic standings, highlighting that benefits go beyond financial aid or technology transfer; exchange of ideas is paramount. This diplomacy can be metaphorically compared to a vehicle driven at moderate speed, implying that hurried negotiations might not yield sustainable outcomes. Economic considerations heavily influence a nation's foreign policy, making competitiveness crucial to ensure mutual benefits. However, the effectiveness of economic diplomacy can be compromised if there exists political animosity between the countries involved. A stable political regime is vital for fostering predictability and certainty in foreign policies, promoting stronger economic ties with other nations.

The intricacies of 'bargaining power' are central to economic diplomacy. Despite misconceptions, negotiation isn't a display of 'muscle power' but rather an intellectual exercise that demands a comprehensive understanding of the subject matter. Moreover, the capacity for meaningful negotiation is underpinned by diplomacy's participatory nature, mandating informed participants with a broad perspective and an appreciation for mutual interests. This necessitates training diplomats in diverse disciplines, keeping them abreast of contemporary economic and politico-economic issues. Notably, the global landscape highlights the interconnectedness of economic and general diplomacy. A country's economic strength not only ensures its sustainability but also amplifies its voice in the international arena, underscoring the symbiotic relationship between economic relations and broader diplomatic engagements.

2.7 Theoretical Aspects of Economic Diplomacy: A Comprehensive Analysis

Economic diplomacy, as an integral dimension of international relations, revolves around the utilization of economic resources and tools to achieve foreign policy objectives. The theoretical underpinnings of economic diplomacy are multifaceted, with several schools of thought influencing its understanding and application.

Realism and Mercantilism: Historically rooted in mercantilism, realists view economic diplomacy as a zero-sum game where states act based on their national interests, often at the expense of others (Morgenthau 1948). In this perspective, economic strength and national security are intrinsically linked. States seek relative gains in trade, and economic tools like sanctions or embargoes are instruments of power politics.

Liberalism: Contrary to realists, liberals believe in the possibility of mutual benefits through economic cooperation (Keohane and Nye 1977). They argue that economic interdependence reduces the likelihood of conflict, emphasizing the role of international institutions in facilitating economic diplomacy. For liberals, international economic collaboration, through mechanisms like free trade agreements, enhances global welfare.

Constructivism: Constructivists focus on the role of ideas, beliefs, and norms in shaping economic diplomacy (Wendt 1992). They argue that state interests are socially constructed and not just material. Economic relations are also influenced by shared ideas, such as the global acceptance of market principles or sustainable development.

Institutionalism: Institutional theorists underscore the significance of international organizations in economic diplomacy (Keohane 1984). They posit that institutions reduce uncertainties, establish information-sharing mechanisms, and provide platforms for states to cooperate on economic matters. Organizations like the World Trade Organization or the International Monetary Fund exemplify how institutions facilitate economic diplomacy.

Complex Interdependence: Keohane and Nye's (1977) theory of complex interdependence argues that in the modern world, states are connected through multiple channels, not just military. Economic links play a vital role in diplomacy, and the line between domestic and international issues blurs. Economic diplomacy in this view becomes a tool for managing these intricate interdependencies.

Game Theory: This mathematical approach studies strategic interactions among rational actors. In economic diplomacy, game theory might be used to analyze negotiation strategies or the potential outcomes of trade wars.

In sum, economic diplomacy operates at the confluence of economic interests and foreign policy objectives. While realists highlight power dynamics and relative gains, liberals and institutionalists emphasize mutual benefits and the role of international institutions. Constructivists, on the other hand, bring attention to the ideational aspects influencing economic interactions. The richness of these theoretical perspectives underscores the multidimensionality of economic diplomacy in the contemporary world order.

2.7.1 Instruments of Economic Diplomacy

Economic diplomacy encompasses the strategies and tools used by states and non-state actors to achieve economic goals within the realm of foreign relations. Various instruments facilitate these objectives, encompassing both cooperative and coercive mechanisms.

1. **Trade Agreements**: Bilateral, regional, and multilateral trade agreements are pivotal tools in economic diplomacy. They can open markets, reduce trade barriers, and set rules for international trade.
2. **Economic Sanctions**: As coercive tools, sanctions aim to change behavior by inflicting economic pain. They can be comprehensive, targeting an entire economy, or targeted, focusing on specific sectors or individuals.
3. **Foreign Aid and Development Assistance**: Countries provide financial and technical assistance to foster development, gain influence, or support allies. Aid can be bilateral or multilateral through institutions like the World Bank or United Nations.
4. **Investment Treaties**: Bilateral Investment Treaties (BITs) protect and promote foreign investments by establishing rules and dispute settlement mechanisms.
5. **Trade Missions and Promotion**: Governments organize trade missions, fairs, and expos to promote domestic industries abroad and attract foreign investment.
6. **Currency and Exchange Rate Policies**: By manipulating exchange rates, governments can influence trade balances, with a devalued currency making exports cheaper and imports more expensive.
7. **Export Credits and Guarantees**: Governments can offer credits and guarantees to mitigate risks for domestic companies doing business abroad, facilitating international trade.
8. **Technical Standards and Regulations**: States can use standards and regulations as non-tariff barriers, protecting domestic industries or ensuring product quality. However, these can also become obstacles to trade if not harmonized internationally.
9. **Economic Espionage**: Some governments engage in or tacitly support the covert collection of sensitive economic information to benefit domestic industries.
10. **Cultural and Educational Exchanges**: Soft power instruments, these programs foster mutual understanding and build networks, indirectly supporting economic objectives (Nye 2008).

In essence, the instruments of economic diplomacy range from cooperative mechanisms like trade agreements and development assistance to more confrontational tools like sanctions. Their application depends on the specific goals and context of the diplomatic engagement.

2.7.2 Modern Challenges and Evolution of Economic Diplomacy

There are several challenges of economic diplomacy in the digital age, particularly with the rise of e-commerce, digital currencies, and cyber threats. The traditional tools of economic diplomacy may need adaptation to remain effective. Economic diplomacy has undergone substantial transformation in response to global shifts, with modern challenges shaping its evolution.

The rise of emerging economies like the BRICS (Brazil, Russia, India, China, and South Africa) has redistributed economic power, prompting shifts in trade relations and investment patterns. Globalization has amplified interdependence among nations, making unilateral economic decisions more consequential for global stability. For instance, the U.S.-China trade war has impacted global supply chains and markets. Digitalization presents another frontier. Cybersecurity, e-commerce rules, and digital taxation have become pivotal in economic diplomacy, reflecting the increasing significance of the digital economy. Climate change introduces additional complexity. Economic agreements now often include environmental stipulations, and there's a push for sustainable investments and green technologies in diplomatic engagements.

Lastly, the COVID-19 pandemic underscored the need for international collaboration in public health as an economic imperative. Supply chain disruptions and debates over vaccine distribution showcased the intertwined nature of health and economic diplomacy. In essence, economic diplomacy's evolution is a response to the multifaceted challenges of a rapidly changing global landscape.

2.8 Role of Economic Diplomacy in International Trade and Economic Relations

Economic diplomacy can be described as the use of diplomatic strategies to influence economic activities and dynamics on a global scale. It operates at the nexus of international trade, investment, finance, and policymaking. Economic diplomacy represents a synergy between traditional diplomatic activities and the imperatives of global economic exchanges, fostering both competition and cooperation among states.

1. **Promotion of Trade and Investments**

One of the primary roles of economic diplomacy is the promotion of trade and investments. Through activities such as bilateral talks, trade fairs, and investment seminars, countries use diplomacy to advertise their economic potential. Diplomatic missions have trade and investment wings that interact with the business community and policymakers of the host countries to create opportunities for domestic businesses. For instance, bilateral investment treaties and free trade agreements, which are often a product of extensive diplomatic negotiations, can pave the way for improved trade and investment flows.

2. **Crisis Management and Economic Stabilization**

During financial crises, economic diplomacy is of vital importance. The G20 meetings, which are a form of multilateral economic diplomacy, played a crucial role in coordinating responses to the 2008 financial crisis (Helleiner 2010). Economic diplomacy ensures that countries maintain open channels of communication during such times, thereby preventing protectionist responses that could worsen the situation.

3. **Advocacy for Global Economic Governance**

Global economic governance, which refers to the establishment of international rules for economic interaction, is another domain shaped by economic diplomacy. Organizations like the World Trade Organization (WTO), International Monetary Fund (IMF), and World Bank require consensus-building, which is achieved through economic diplomatic engagements. Through such diplomatic endeavors, countries shape global economic norms, rules, and policies in line with their national interests.

4. **Soft Power and Economic Influence**

Economic diplomacy also plays into the narrative of soft power. Nations often provide economic assistance, preferential trade agreements, or infrastructure investments to other countries as a means of gaining influence without using coercive measures. The Belt and Road Initiative by China is a testament to how economic tools can be employed in diplomacy to expand influence and foster closer ties with other nations.

5. **Resolving Trade Disputes**

International economic relations cannot exist without trade disputes. Instead of escalating into full-scale economic warfare, economic diplomacy provides avenues for conflict resolution. The WTO's Dispute Settlement Mechanism, for instance, allows countries to file complaints and pursue diplomatic resolution, thereby ensuring the stability and predictability of the global trade system (Busch and Reinhardt 2003).

6. **Encouraging Sustainable Development**

The accordance of sustainable development objectives has also incorporated economic diplomacy. Diplomatic channels facilitate cooperation on global issues such as climate change, ensuring that economic development does not compromise environmental sustainability. The Paris Agreement of 2015 is an illustration of how diplomatic efforts can lead to a collective commitment to a sustainable economic future.

7. **Facilitation of Foreign Direct Investment (FDI)**

Economic diplomacy aides in attracting foreign investments beyond traditional trade. Diplomatic efforts guarantee that potential investors are well-informed about investment opportunities and that the business climate is favorable.

8. **Gathering Trade Intelligence**

When it comes to international trade, economic diplomats act as a country's ears and eyes. They are crucial because they inform policymakers and businesses about market trends, opportunities, difficulties, and potential threats.

9. **Crisis Management**

Economic diplomacy allows for communication, coordination, and the pooling of resources in times of economic crisis. These kinds of cooperation procedures can stop problems from getting worse and hurting international commercial relations.

International commerce and economic ties continue to rely heavily on economic diplomacy, even in today's globalized world where economic interdependence is the norm. Economic diplomacy promotes peace and prosperity on the global economy by easing barriers to trade, investment, and mutual economic understanding. However, the dynamic nature of international power and economy necessitates the constant modification and improvement of economic diplomacy instruments and approaches.

2.9 Developing Countries and Economic Diplomacy

Economic diplomacy, which encompasses the strategies governments employ to foster their national economic interests in the international arena, holds specific significance for developing countries. These nations often grapple with challenges including limited resources, economic vulnerabilities, and a quest for sustainable growth. Engaging effectively in economic diplomacy can play a transformative role for such states. This essay delves into the intricate relationship between developing countries and economic diplomacy.

1. **Trade Agreements and Market Access**

For developing countries, economic diplomacy offers avenues to negotiate favorable terms in bilateral and multilateral trade agreements. Such nations often seek preferential market access to developed countries' markets, aiming to boost exports. The Generalized System of Preferences (GSP) granted by various developed countries is a testament to successful economic diplomacy by developing nations.

2. **Foreign Direct Investment (FDI) Attraction**

Developing countries, in their bid to accelerate economic growth, often look to attract foreign investments. Economic diplomacy aids in showcasing the investment potential, ensuring investor protections, and creating conducive environments for FDI. Initiatives like Investment Promotion Agencies (IPAs) and investment treaties serve this purpose.

3. **South-South Cooperation**

Many developing countries have increasingly looked towards South-South cooperation as an avenue for mutual economic growth. This form of collaboration, promoted through economic diplomacy, involves trade, investments, and technological exchanges between developing countries. The IBSA Dialogue Forum, encompassing India, Brazil, and South Africa, is a notable example.

4. **Access to Technology and Skills**

Developing nations, through economic diplomacy, can negotiate technology transfers and skill-building partnerships. Such transfers are vital for their industrial growth, infrastructure development, and capacity-building.

5. **Advocacy in International Economic Institutions**

Developing countries often face disadvantages in international economic forums like the WTO, IMF, and World Bank, where decisions might be skewed in favor of the developed world. Through collective economic diplomacy, as seen in blocs like the G77, these nations can amplify their voice and advocate for their interests.

6. **Addressing Economic Asymmetries**

Developing nations can rectify imbalances in the global economic order through economic diplomacy. Together, they may demand a more just global economy through coalition building and collective bargaining.

7. **Tapping into the Diaspora**

Engaging with their diaspora through economic diplomacy can turn them into valuable resources for developing nations. Among the many dimensions of this dynamic are remittances, transfers of expertise, and lobbying efforts by the diaspora.

8. **Debt Relief and Financial Assistance**

Developing countries often grapple with enormous external debts which stifle their economic growth. Through economic diplomacy, they can negotiate for debt relief, re-scheduling, or more favorable lending terms with creditors and international financial institutions. The Heavily Indebted Poor Countries (HIPC) initiative, guided by the World Bank and IMF, exemplifies this dimension of economic diplomacy, aiming to ensure that poor countries do not face debt burdens they cannot manage.

9. **Commodities and Pricing**

Many developing countries heavily depend on commodities exports. Price volatility in commodities can significantly affect their economies. Through economic diplomacy, these countries can engage in dialogues, form cartels, or seek strategic partnerships to stabilize prices. The formation of the Organization of Petroleum Exporting Countries (OPEC) is a prime example of such a diplomatic endeavor to influence petroleum prices.

10. **Climate Finance and Environmental Diplomacy**

With the growing impacts of climate change, developing countries, particularly those vulnerable to its effects, employ economic diplomacy to secure financial, technological, and capacity-building support. The negotiations leading to the Green Climate Fund, meant to channel resources from the developed to the developing world, underscores this point.

11. **Enhancing Regional Integration**

Regional integration can offer economies of scale, investment opportunities, and a united front in global negotiations. Economic diplomacy aids in the formulation and smooth functioning of regional economic communities. The establishment of the Association of Southeast Asian Nations (ASEAN) and its economic community is illustrative of this rationale.

12. **Non-Tariff Barriers and Standards Negotiation**

While tariffs remain a concern, non-tariff barriers such as health standards, quality checks, and certifications are increasingly becoming significant impediments to trade. Developing countries use economic diplomacy to ensure that these standards don't arbitrarily stifle their exports. The "SPS (Sanitary and Phytosanitary) and TBT (Technical Barriers to Trade) Agreements" within the WTO framework highlight these concerns and the diplomatic channels available to address them (Hoekman and Kostecki 2009).

For developing countries, economic diplomacy is not just a tool but a necessity. It allows them to navigate the complexities of the global economic system, ensuring that their developmental needs are addressed. By fostering strategic partnerships, attracting investments, and ensuring a favorable global economic policy environment, economic diplomacy can significantly catalyze the developmental aspirations of these nations.

2.9.1 Developing Countries and Economic Diplomacy: A New Landscape

There's a prevailing misconception that developing countries hold a minor role in international economic diplomacy. Such a view undermines the potential and past contributions of these nations. True diplomacy, be it economic or otherwise, hinges on negotiations that are mutually beneficial. Dominance-based decisions are more prone to violation, as showcased by the push for a New International Economic Order in 1974, a movement against the Bretton Woods setup which largely ignored developing countries.

It's pivotal that international economic diplomacy upholds principles of non-discrimination and fairness, ensuring the interests of the most vulnerable countries are considered, regardless of their geographic or economic status. Misunderstandings about "bargaining power", which is the essence of genuine negotiation, have often been manipulated, potentially sidelining the voices of developing nations.

However, the contributions of these countries are undeniable. Examples include the UN Resolution on Permanent Sovereignty over Natural Resources in 1962 and the concept of the Exclusive Economic Zone in the UN Convention on the Law of the Sea. Additionally, the Charter of Economic Rights and Duties of States 1974 presents foundational principles for equitable international economic diplomacy. With nations like Brazil, China, and India becoming global economic powerhouses, the dynamics are shifting. Recognizing and nurturing the potential of developing nations is vital for a balanced global economic environment.

As the developed world seeks expanded markets, international cooperation becomes indispensable. Economic diplomacy is broad, covering trade, banking, finance, investments, and treaty negotiations. Bilateral investment treaties, rooted in international law principles, signify economic diplomacy's real-world applications. Such treaties often involve developed-developing or advanced developing-developing country partnerships. It's essential to comprehend how these treaties reflect the parties' economic policies and to appreciate the role of state contracts in fostering trust among foreign investors.

2.9.2 The Role of Non-governmental Organizations in Economic Diplomacy

Economic diplomacy encompasses a myriad of activities related to cross-border economic endeavors, encompassing both state and non-state actors (Baine and Woolcock 2003). At its core, state-led economic diplomacy can be categorized into three primary areas:

Bilateral Economic Diplomacy: This pertains to leveraging political influence to bolster international trade and investment, optimize market functionality, and mitigate risks of transnational transactions. Examples include export promotion and state visits.

Strengthening Economic Security: Here, economic assets are strategically used to amplify the costs of conflicts while reinforcing the benefits of collaboration and stable political ties. Such strategies may include trade agreements or even political trade distortions like embargoes.

Trade Diplomacy: This target fosters a conducive political and economic global environment. Organizations like the WTO, OECD, and EU play pivotal roles in this dimension, focusing on multilateral negotiations. However, the potential benefits of such negotiations, as evidenced in discussions on the Uruguay and Doha Rounds, are well-acknowledged and require minimal elaboration.

Crucially, Baine and Woolcock's definition underscores the evolving role of non-state actors in economic diplomacy. As the influence of entities like multinational enterprises (MNEs), consumer groups, and non-governmental organizations (NGOs) burgeons, the dynamics of international diplomacy shift significantly. The involvement of these non-state actors delineates a clear departure from traditional diplomatic frameworks and highlights a modern, multifaceted international ecosystem (Saner and Yiu 2003).

2.9.3 Economic Diplomacy and Commercial Diplomacy: An Overview

Economic and commercial diplomacy serve distinct yet interrelated roles in the realm of international relations. At its core, economic diplomacy focuses on policymaking, aimed at bolstering a nation's economic security through diplomatic channels. This encompasses both bilateral relations, such as securing foreign investments or technology transfers, and multilateral discussions within global organizations or regional bodies, where the emphasis is on shaping new international economic norms and rules. Commercial diplomacy, though narrower in scope, zeroes in on transactional aspects like import–export deals, trade preferences, or other business-related ventures between states. It often operates on a bilateral or regional level, and its primary aim is to boost business development and profit-making rather than policy formulation. This sector predominantly sees the involvement of commercial counselors or trade representatives whose chief goal is the promotion of business between their home country and a host nation.

While economic diplomacy can be seen as a broader policy-based approach, characterized by mutual benefits and guided by international frameworks like the Charter of Economic Rights and Duties of States of 1974 and the Doha Declaration of 2001, commercial diplomacy adopts a more tactical stance. Its essence is the generation of profits, often executed under models like the buy-back system. It's worth noting that commercial diplomacy sometimes occupies a secondary status, with certain missions outsourcing commercial activities, potentially due to a lack of negotiation expertise or other constraints. This can be counterproductive since commercial diplomacy significantly influences a country's foreign policy. In conclusion, both economic and commercial diplomacy play pivotal roles in shaping a nation's international stance. The balance between these two facets is crucial, as they collectively determine a country's foreign mission size and direction. Effective negotiations and capacity-building remain essential to ensure developing countries don't slide into a diplomacy of dependency.

2.9.4 Bilateral Relations

Bilateral relations in international politics refer to the interactions, agreements, and negotiations between two specific sovereign countries. Unlike multilateral relations, which involve multiple countries, bilateral relations focus on the relationship between two states. This can encompass various areas such as trade, defense, diplomatic ties, cultural exchanges, and more. The significance of diplomatic connections between nations has its origins in the long and storied history of bilateral partnerships. Countries have been actively pursuing connections with other governments to defend and advance their interests ever since the Treaty of Westphalia in 1648. Such connections are the backbone of diplomacy, which is why many nations set up diplomatic missions or consulates abroad. Bilateral trade agreements are a good illustration of these types of ties in operation. Such deals can have far-reaching effects, affecting the global economy and the distribution of commodities and services. In terms of defense and strategic alliances, bilateral ties are also crucial. One such example is the U.S.-Japan Security Treaty of 1960, which shows how two countries can work together for their own security.

Historical circumstances, geopolitical factors, and domestic political scenarios can all have a significant impact on the dynamics of bilateral ties. Cooperation is possible in some relationships, but conflict is also possible. Even in today's rapidly globalizing world, the importance of bilateral interactions is clear in international politics. In the field of international relations and diplomacy, bilateral connections are a well-studied issue despite the rise of multilateral forums and regional organizations. Here are some definitions and citations from some of the most respected minds in the field:

Joseph S. Nye, Jr.:

> Bilateral relations involve two nations. Similar to interpersonal relationships, they can range from affectionate to hostile. However, nations are not individuals, and their relationships are impersonal. They are complex organizations that represent a variety of interest groups and whose internal processes and domestic environments influence them.

Reference: Nye, J.S. (2008). International Relations: The Most Important Relationship. In The Powers to Lead. Oxford University Press.

Henry Kissinger:

> Bilateral relations are the political, economic, cultural, or social ties between two sovereign nations. It is a complex ballet of mutual interests, balancing cooperation and conflict.

Reference: Kissinger, H. (1994). Diplomacy. Simon & Schuster.

John Mearsheimer:

> 'The framework of the international system includes bilateral connections. The greater configuration of power in the system is another factor that could influence the degree of interaction between two states."

Reference: Mearsheimer, J.J. (2001). The Tragedy of Great Power Politics. W.W. Norton & Company.

Robert Keohane:

> 'Bilateral contacts typically play a crucial role in an integrated global system. They are shaped not only by bilateral interactions between nations but also by broader institutional frameworks and networks of multilateral ties.'

Reference: Keohane, R.O. (1984). After Hegemony: Cooperation and Discord in the World Political Economy. Princeton University Press.

2.9.5 *History of Bilateral Relations: A Brief Overview*

Since ancient times, engagements between two independent states—or bilateral ties—have served as the cornerstone of international relations. In ancient times, Egypt and Mesopotamia sent envoys to each other to discuss matters of peace, trade, and alliance. The Treaty of Tordesillas of 1494, for example, was a treaty between Spain and Portugal that established boundaries for each country's colonial sphere of influence. The Westphalian system formalized the concept of state sovereignty in 1648, elevating bilateral relations to the forefront of international politics.

The nineteenth century saw the rise of 'Great Power' diplomacy. Major European powers used bilateral treaties to maintain the balance of power, as evident in the Treaty of Tilsit of 1807 between France and Russia or the numerous accords between Germany and Russia. Post-World War II, the Cold War era was dominated by the U.S.-Soviet bilateral relationship, where both superpowers engaged in a complex dance of cooperation and confrontation. At the same time, newly independent nations from Africa and Asia forged bilateral ties as a part of the Non-Aligned Movement. In recent times, globalization has intensified bilateral trade and investment relations. While multilateral forums like the UN remain significant, bilateralism has regained importance, as seen in U.S.-China trade negotiations or India-Russia defense deals. Bilateral relations have, thus, been a consistent and evolving feature of international diplomacy, shaping the contours of global politics across ages.

2.9.6 *Bilateral Relations as the Foundations of International Affairs*

Bilateral relations stand as one of the cornerstones in the architecture of international affairs. These relations, established between two countries, form the microcosm from which larger regional and global trends often emerge. Let's delve deeper into their foundational importance:

1. **Historical Legacy**: Bilateral interactions were the norm in early international relations. Countries historically interacted with one another in pairs, laying the groundwork for the conventions and principles that guide current international relations through treaties, trade routes, and conflicts.
2. **Trade and Economic Growth**: For a long time now, bilateral trade agreements have been a driving force behind increased prosperity. These pacts have helped shape policies, establish trade norms, and increase economic interdependence, all of which serve to reduce the likelihood of war.
3. **Security and Strategic Interests**: An integral aspect of the dynamics of international security are bilateral defense pacts, intelligence sharing agreements, and military collaborations. The nature of these bilateral ties can affect the safety of neighboring countries.

4. **Cultural and Soft Power Dynamics**: To promote soft power, shape public perception, and often determine diplomatic outcomes, bilateral cultural exchanges and agreements are essential.
5. **Platform for Multilateralism**: Bilateral relations often act as stepping stones towards broader multilateral engagements. For instance, two countries with strong bilateral ties may advocate for shared interests in larger international forums.
6. **Conflict Resolution**: Bilateral negotiations and dialogues have been instrumental in peacefully resolving disputes, setting the stage for larger regional peace initiatives.

In summary, while multilateral engagements and international institutions are undeniably significant, the edifice of international affairs rests heavily on the foundational blocks of bilateral relations.

2.9.7 Scope and Functions of Bilateral Relations

1. **Trade and Commerce**: Bilateral relations often revolve around trade agreements, encompassing tariffs, trade barriers, and the establishment of free trade zones. These interactions determine the flow of goods and services between the two countries in question.
2. **Defense and Security**: Bilateral defense ties can include joint military exercises, arms sales, defense pacts, and intelligence sharing. Such engagements aim to bolster mutual security and strategic interests.
3. **Cultural and Educational Exchanges**: Beyond politics and economics, countries engage in the mutual promotion of culture, language, art, and academic pursuits. Such exchanges facilitate deeper understanding and strengthen people-to-people ties.
4. **Diplomatic Engagements**: Diplomatic discussions, negotiations, and frequent high-level visits are all part of this process since they help spread positive feelings and increase mutual understanding between countries.
5. **Environmental and Energy Cooperation**: Countries increasingly work together on topics like fighting climate change, developing renewable energy sources, and implementing conservation programs as a result of globalization.
6. **Health and Scientific Collaboration**: Countries work together in research, innovation, and health initiatives due to the need to address global health issues and technical gaps (Fidler 2001).
7. **Migration and Labor**: There are many bilateral labor agreements in place, allowing for the controlled migration of workers between countries. Such pacts can help alleviate employment difficulties in one nation while supplying needed workers in another (Martin 2004).

8. **Counterterrorism**: When confronted with international terrorism, several nations build bilateral ties in order to share information, carry out joint operations, and plan effective counterterrorism strategies.
9. **Infrastructure Development**: In order to maximize the benefits for both countries, bilateral agreements are often used to support investments in infrastructure including roads, trains, and ports.
10. **Agricultural Collaboration**: In order to ease the interchange of agricultural goods, innovations, and best practices, countries with complementary agricultural capabilities and requirements might enter into agreements with one another.

2.9.8 Functions

1. **Facilitate Dialogue**: The established framework of bilateral relations allows for the open and productive discussion of a wide range of issues between nations.
2. **Promote Economic Growth**: Investment, the sharing of technologies, and expanded trade are all possible results of bilateral trade agreements.
3. **Ensure Mutual Security**: Defense and strategic partnerships can deter common threats, ensure regional stability, and allow for resource pooling.
4. **Forge Alliances**: Bilateral interactions can lead to deeper alliances, shaping regional and global power dynamics.
5. **Conflict Resolution**: Bilateral dialogues can aid in the peaceful resolution of disputes, avoiding potential escalations.
6. **Soft Power Promotion**: Cultural, educational, and technological exchanges enhance a nation's soft power, influencing foreign publics and governments.
7. **Disaster Response and Humanitarian Aid**: Bilateral ties often get activated during natural disasters, where one country can offer immediate relief, resources, and expertise to another in need.
8. **Addressing Transnational Crimes**: Bilateral cooperation can be essential in combating transnational crimes like drug trafficking, human smuggling, and cybercrimes.
9. **Economic Stability**: Bilateral currency swap agreements can be crucial during economic crises, helping stabilize national currencies and fortify economic ties.
10. **Technology and Innovation**: Bilateral relations often encompass technological collaborations in sectors like IT, space research, and clean energy, spurring innovation and shared growth.

2.9.9 Role of Economic Diplomacy in Bilateral Relations

Economic diplomacy refers to the use of political influence by states to secure economic benefits, leveraging policy, negotiations, and agreements related to trade, investment, aid, and other economic elements. In the realm of bilateral relations, economic diplomacy plays a pivotal role in shaping the dynamics between two nations.

Trade Agreements and Market Access: Central to economic diplomacy is the negotiation of trade agreements. Whether comprehensive or sector-specific, these agreements determine tariffs, quotas, and regulations that dictate the flow of goods and services between two nations. Through effective economic diplomacy, nations can secure preferential trade terms, boost exports, and ensure balanced trade relations. For instance, the U.S.-China Phase One trade agreement of 2020 illustrates how intensive economic diplomacy can lead to agreements even amidst heightened geopolitical tensions.

Investment Promotion and Protection: Bilateral investment treaties (BITs) are foundational instruments of economic diplomacy. They provide a framework for investments between countries, offering protections against expropriation, ensuring fair treatment, and establishing mechanisms for dispute resolution. Such treaties bolster investor confidence and increase Foreign Direct Investment (FDI) flows. Notably, the BIT between Germany and Pakistan in 1959 marked the onset of such treaties, emphasizing their importance in international economic relations.

Technology and Knowledge Transfer: Economic diplomacy fosters technological collaborations and knowledge exchanges. By cultivating strong bilateral relations, countries can access and share technologies, best practices, and innovations, especially pertinent in sectors like energy, IT, and manufacturing. The Indo-Israel agricultural cooperation, involving the transfer of advanced agricultural technologies to India, is an exemplar of this facet of economic diplomacy.

Resource Diplomacy: For nations lacking essential resources, economic diplomacy facilitates access to crucial commodities like oil, minerals, or water through bilateral engagements. A classic example would be Japan's diplomacy with Middle Eastern nations to ensure a steady supply of oil, vital for its energy-dependent economy.

Economic Aid and Assistance: Economic diplomacy often involves the provision or receipt of economic aid, grants, or concessional loans. Countries use aid as a tool to foster goodwill, secure strategic alliances, or promote development in partner nations. The U.S. Marshall Plan, which aimed to rebuild postwar Europe, exemplifies economic diplomacy's broader strategic and humanitarian goals.

Soft Power Augmentation: Economic partnerships, cultural exchanges, and academic collaborations, rooted in economic diplomacy, enhance a nation's soft power. These engagements influence perceptions, fostering a favorable environment for broader bilateral relations.

References

Bayne N (2008) Financial diplomacy and the credit crunch: the rise of central banks. J Int Aff, 62(1):1–16. https://www.questia.com/library/journal/1G1-189796226/financial-diplomacy-and-the-credit-crunch-the-rise, http://www.jstor.org/stable/24358141

Busch ML, Reinhardt E (2003) Developing countries and general agreement on tariffs and trade/world trade organization dispute settlement. J World Trade, 3(4):719–735. https://kluwerlawonline.com/journalarticle/Journal+of+World+Trade/37.4/TRAD2003034

Cummings M (2003) Cultural diplomacy and the united states government: a survey. Center for Arts and Culture. https://www.americansforthearts.org/sites/default/files/MCCpaper.pdf

Drake WJ, Nicolaïdis K (1992) Ideas, interests, and institutionalization: "trade in services" and the Uruguay Round, International Organization, 46(1):37–100. https://doi.org/10.1017/S0020818300001454

Goldstein MA, Hotchkiss ES, Sirri ER (2006) Transparency and liquidity: a controlled experiment on corporate bonds. Review of Financial Studies. Oxford University Press (OUP). Available at: https://doi.org/10.1093/rfs/hhl020

Hoekman BM, Kostecki MM (2009) The political economy of the World Trading System, Oxford University Press, Oxford. https://doi.org/10.1093/oso/9780199553761.001.0001

Keohane R, Nye J (1977) Power and interdependence: world politics in transition. Little Brown

Keohane RO (1984) After Hegemony: cooperation and discord in the world political economy. Princeton University Press, Princeton

Kirton JJ (2013) G20 Governance for a Globalized World (1st ed). Routledge. https://doi.org/10.4324/9781315583747

Lee D, Hudson D (2004) The old and new significance of political economy in diplomacy. Review of International Studies, 30(3):343–360. https://doi.org/10.1017/S0260210504006102

Melissen J (2005) The new public diplomacy. Palgrave Macmillan UK eBooks. https://doi.org/10.1057/9780230554931

Morgenthau HJ (1948) Politics among nations. The struggle for power and peace. New York, Knopf

Nye JS (2008) International relations: the most important relationship. In: The powers to lead. Oxford University Press, Oxford

Rana KS (2007) Economic diplomacy: the experience of developing countries. In: The new economic diplomacy: decision-making and negotiations in international economic relations. Ashgate, Aldershot

Reuveny R, Thompson WR (2001) Leading sectors, lead economies, and economic growth. Rev Int Polit Econ 8(4):689–719. https://doi.org/10.1080/09692290110077629

Saner R, Yiu L (2003) International economic diplomacy: Mutations in post-modern times

Scott J, Wilkinson R (2011) China and the WTO. Indiana University Research Center for Chinese Politics & Business Working Paper No. 5, Available at SSRN: https://ssrn.com/abstract=2169307

Sornarajah M (2010) The international law on foreign investment (3rd ed). Cambridge, Cambridge University Press

UNCTAD (2019) World investment report 2019: special economic zones. United Nations Publication

Wendt A (1992) Anarchy is what states make of it: the social construction of power politics, International Organization, 46(2):391–425. https://doi.org/10.1017/S0020818300027764

Chapter 3
Latin American Countries (LAC): An Economic Overview

Abstract This chapter intricately examines the region's economic trajectory, underpinned by its colonial legacy, political transformations, and integration into global markets. Latin America's development is framed by a colonial extractive model that entrenched economic dependencies and systemic inequalities. These foundational disparities persisted post-independence in the nineteenth century when the region, heavily reliant on raw material exports, became a supplier to industrialized nations, notably Britain. Despite political independence, economic autonomy eluded most nations, embedding vulnerabilities that shaped their subsequent histories. The twentieth century introduced diverse economic experiments amidst political volatility, including populist regimes, military dictatorships, and varied economic strategies such as Cuba's socialism and Chile's neoliberal reforms. This period also saw the rise of the Washington Consensus, promoting market liberalization and fiscal discipline, but often criticized for exacerbating inequalities. The chapter captures how these reforms, while integrating LAC economies into global markets, left socio-economic disparities unresolved. The "Pink Tide" of the twenty-first century brought left-leaning governments advocating redistributive policies and social investments, although these efforts faced challenges from fluctuating commodity prices and governance crises. Regional economic diplomacy emerges as a cornerstone of Latin America's modernization, exemplified by the creation of MERCOSUR, the Pacific Alliance, and engagement in trade agreements like NAFTA and the EU-MERCOSUR accord. Case studies of Brazil and Argentina underscore their pivotal roles in regional leadership and bilateral partnerships, including strategic alignments with China, whose trade with LAC grew from $10 billion in 2000 to over $300 billion by 2019. This dynamic, driven by China's demand for raw materials and infrastructure investments, juxtaposes with traditional ties to the U.S. and Europe, reflecting a complex balance of economic dependencies. The chapter also elaborates on LAC's diverse economic systems and resource endowments. Key sectors include resource extraction, such as Venezuela's oil and Chile's copper, agricultural exports like Brazil's soybeans, and emerging industries in technology and tourism. While LAC countries benefit from abundant natural resources, the dependency on commodity exports exposes them to global market fluctuations. The analysis of macroeconomic indicators reveals stark contrasts in GDP growth, inflation, and unemployment rates,

M. T. Islam and M. B. Hossen, *Economic Diplomacy*, The Latin American Studies Book Series, https://doi.org/10.1007/978-3-031-81362-7_3

highlighting economic disparities across the region. For instance, Brazil and Mexico demonstrate economic resilience with diversified industries, while Venezuela grapples with hyperinflation and political instability. Case studies of Brazil and Argentina illustrate the nuanced approaches to economic diplomacy, emphasizing regional integration, South-South cooperation, and strategic global engagements. The chapter also underscores the importance of bilateral and multilateral relations in addressing trade imbalances, resource management, and geopolitical tensions. The growing influence of China in the region, juxtaposed with traditional Western alliances, epitomizes the evolving global economic landscape. Trade dynamics are enriched by regional integration efforts and external partnerships. Bilateral and multilateral agreements, including the Belt and Road Initiative, reshape LAC's position in global supply chains. Challenges persist in balancing growth, equity, and sustainability, with environmental degradation, informal economies, and governance deficits posing ongoing obstacles. This chapter offers a sophisticated lens on LAC's economic evolution, emphasizing its dual struggle between external dependencies and internal reforms. As Latin America navigates the pressures of globalization and its pursuit of equitable development, understanding the interplay of history, policy, and global integration becomes critical for shaping its future trajectory. The region's ability to foster innovation, sustainable practices, and inclusive growth will determine its standing in an increasingly multipolar world.

Keywords LAC · MERCOSUR · Pacific alliance · EU · Bilateral and multilateral relation

Latin America's history, marked by the interplay of politics and economics, encompasses the aftermath of colonization, economic dependencies, and a myriad of political shifts, profoundly impacting its economic trajectory. While the nineteenth century ushered in independence, it did not necessarily equate to economic autonomy, with foreign powers like Britain positioning Latin America as a supplier of raw materials. The twentieth century saw a rise in populism, with figures like Perón emphasizing growth through export revenues. However, military coups and varying economic strategies from Chile's neoliberalism to Cuba's socialism became dominant in the latter half (Hershberg and Rosen 2006). The 'Washington Consensus' of the late twentieth century, emphasizing market reforms, became influential yet contentious (Williamson 1990). The twenty-first century's 'Pink Tide', representing the ascendancy of left-leaning governments, highlighted the continuous quest for an economic model prioritizing both growth and equity.

3.1 History of Latin America in the Context of Political Economy

Latin America, comprising diverse nations, has undergone a rich political and economic tapestry shaped by both internal dynamics and external pressures. This intricate history has seen the region grapple with the repercussions of colonization, economic shifts, and political transformations, all of which have profoundly influenced its economic trajectory.

Colonial Legacy (Sixteenth to Nineteenth Century)

European colonization, primarily by Spain and Portugal, deeply affected Latin America's political economy. The encomienda system, an arrangement where colonists had the right to demand labor and tribute from indigenous communities, sowed the seeds of societal inequality and skewed wealth distribution. Furthermore, an economic pattern was set with the emphasis on extracting valuable resources, epitomized by the silver mining activities in Potosí.

Nineteenth Century—Independence and Economic Openness

With the decline of Spanish and Portuguese empires, Latin America saw the emergence of independent republics. Yet, economic independence remained elusive. British industrial influence turned the region into a pivotal supplier of raw materials while increasing its reliance on imported finished goods, perpetuating economic disparities.

Twentieth Century—Populism, Dictatorships, and Economic Experiments

The twentieth century heralded the advent of populist figures, with Argentina's Perón epitomizing a promise of inclusive economic growth, often funded through commodity exports. However, many nations were soon engulfed by military coups, with Cold War politics heavily influencing these transitions. Economic strategies during this era were diverse, from Pinochet's neoliberalism in Chile to Castro's socialist Cuba.

The latter part of the twentieth century saw the 'Washington Consensus' influencing the region's economic policies. Despite advocating market reforms and fiscal discipline, it faced criticism for intensifying inequalities and sidelining social concerns (Williamson 1990).

Twenty-First Century—Left Turns and Economic Uncertainty

The twenty-first century began with the ascendance of leftist governments, exemplified by Venezuela's Chávez and Brazil's Lula. Supported by robust commodity prices, these regimes championed redistributive policies and social investments. Yet, challenges emerged as commodity prices dwindled and corruption scandals surfaced. The political economy of Latin America is a testament to the region's enduring struggle between external impositions and internal aspirations. As the region forges ahead, its quest remains: crafting an economic framework that balances growth with equitable societal progression.

3.2 Background of Latin American Countries at a Glance

A comprehensive overview of Latin America by country would be a monumental endeavor given the depth and breadth of each nation's history, politics, economics, and culture. To keep it concise while providing a meaningful snapshot, I'll give a brief overview for a selection of key countries in the region.

3.2.1 Argentina

History: Several indigenous groups called Argentina home before the Spanish arrived in the early sixteenth century and began colonizing the area. It gained its independence in 1816.

Politics: Renowned for its historical era characterized by military governance and the infamous "Dirty War". In recent times, the political system has undergone a transformation, resulting in the establishment of a stable democratic regime.

Economics: Historically characterized as an agrarian economy, the region has undergone a process of diversification, resulting in the emergence of robust sectors, notably the wine sector.

Culture: Famous for its contributions to the genre of tango music and its captivating dance form. The presence of European immigrants, specifically Italians and Spaniards, is discernible in the cultural landscape.

Source: Rock, D. (1987). Argentina: 1516–1987. University of California Press.

3.2.2 Brazil

History: Portuguese colonization began in the 1500 s, and independence came about in 1822.

Politics: Following a string of military dictatorships, Brazil made the transition to democracy in the 1980s.

Economics: One of the prominent economies on a worldwide scale, characterized by its notable strengths in several sectors like as agriculture, mining, manufacturing, and services.

Culture: Brazil is renowned for its cultural contributions, including the vibrant samba music, the globally celebrated sport of football, and the exuberant funfair festivities. The region exhibits a diverse amalgamation of indigenous, African, Portuguese, and various other cultural influences.

Source: Skidmore, T. E. (1999). Brazil: Five Centuries of Change. Oxford University Press.

3.2.3 Chile

History: The region, which was formerly inhabited by the Inca Empire, underwent colonization by Spain around the mid-sixteenth century. The independence of Chile was achieved in the year 1818.

Politics: Experienced a military coup in 1973 which introduced the authoritarian rule of Pinochet. Returned to democracy in the 1990s.

Economics: South America boasts a robust economy that heavily relies on the exportation of minerals, with a special emphasis on copper.

Culture: Influenced by Spanish and indigenous cultures, with national dance Cueca and the famous poet Pablo Neruda as cultural icons.

Source: Collier, S., & Sater, W. F. (2004). a History of Chile: 1808–2002. Cambridge University Press.

3.2.4 Mexico

History: Home to ancient civilizations like the Maya and Aztecs, it was colonized by Spain in 1521 and gained independence in 1821.

Politics: Dominated by the Institutional Revolutionary Party (PRI) for most of the twentieth century, now a multiparty democracy.

Economics: Diverse economy with agriculture, oil, tourism, and manufacturing sectors.

Culture: Rich in pre-Columbian traditions, with vibrant music (like Mariachi), food, and festivals such as Day of the Dead.

Source: Meyer, M. C., & Sherman, W. L. (2007). the Course of Mexican History. Oxford University Press.

3.2.5 Peru

History: The heartland of the Inca Empire, it became a Spanish colony in the sixteenth century and achieved independence in 1821.

Politics: Has experienced various coups and changes in governance; now a representative democratic republic.

Economics: Mining, particularly silver, copper, and gold, plays a major role. Also noted for its agricultural products.

Culture: Deeply rooted in its Incan heritage, it's known for Quechua language, Andean music, and landmarks like Machu Picchu.

Source: Klaren, P. F. (2000). Peru: Society and Nationhood in the Andes. Oxford University Press.

3.2.6 Colombia

History: Before Spanish colonization in the sixteenth century, Colombia was home to various indigenous cultures. It achieved independence in 1819.

Politics: Experienced internal conflict with guerilla groups like FARC, but peace efforts are ongoing.

Economics: Known for its coffee exports, oil, and emeralds.

Culture: Cumbia and Vallenato are traditional musical genres. Gabriel García Márquez, a Nobel laureate in Literature, hails from here.

Source: Bushnell, D. (1993). the Making of Modern Colombia: A Nation in Spite of Itself. University of California Press.

3.2.7 Ecuador

History: Inhabited by various indigenous groups before Spanish colonization in the 1530 s. Became independent in 1830.

Politics: Historically marked by instabilities; however, it has maintained relative democratic stability in recent decades.

Economics: Oil is a significant export, along with bananas and shrimp.

Culture: Ecuador boasts a rich indigenous heritage, especially in the Andes. The Galápagos Islands are a unique ecological treasure.

Source: Gerlach, a. (2003). Indians, Oil, and Politics: A Recent History of Ecuador. SR Books.

3.2.8 Bolivia

History: The heartland of ancient cultures like the Tiwanaku and part of the Inca Empire, Bolivia was colonized by Spain in the sixteenth century.

Politics: Bolivia has experienced numerous coups. The politics recently have revolved around indigenous rights and resource management.

Economics: Natural gas and mining are crucial. Bolivia is the world's largest producer of quinoa.

Culture: A strong indigenous influence, especially Aymara and Quechua. Celebrated for its traditional music and dances.

Source: Klein, H. S. (2003). a Concise History of Bolivia. Cambridge University Press.

3.2.9 Paraguay

History: Before Spanish colonization in the sixteenth century, it was inhabited by the Guaraní people. Paraguay declared independence in 1811.

Politics: The twentieth century saw a long-standing dictatorship under Alfredo Stroessner.

Economics: Agriculture-driven, with soybeans and beef as significant exports.

Culture: Guaraní culture is influential, with the Guaraní language spoken alongside Spanish. Folk music and dance are integral to Paraguayan identity.

Source: Lewis, P. H. (1982). Paraguay Under Stroessner. University of North Carolina Press.

3.2.10 Uruguay

History: Initially inhabited by the Charrúa people, Uruguay faced Spanish and Portuguese interests until its independence in 1828.

Politics: Historically more stable than many of its neighbors, with a strong democratic tradition.

Economics: Agriculture-centric, particularly beef and wool.

Culture: Tango, like in Argentina, has roots here. Celebrated for its literature and visual arts.

Source: Weinstein, M. (1975). Uruguay: Democracy at the Crossroads. Westview Press.

3.2.11 Venezuela

History: Home to various indigenous groups, it became a Spanish colony in the sixteenth century and gained independence in 1811.

Politics: twenty-first century marked by Hugo Chávez's Bolivarian Revolution and subsequent political and economic challenges.

Economics: Heavily reliant on oil exports, facing economic difficulties in recent years.

Culture: Diverse influences, from indigenous to Afro-Venezuelan. Known for its music, like Joropo.

Source: Cannon, B. (2009). Hugo Chávez and the Bolivarian Revolution: Populism and Democracy in a Globalised Age. Manchester University Press.

3.3 Economic Reformation of Latin American Countries and the Growth of Liberal Democracy

The late twentieth and early twenty-first centuries witnessed a profound transformation in Latin American political and economic landscapes. The economic crises of the 1980s, often referred to as the 'Lost Decade', combined with growing demands for political freedoms, led to sweeping reforms across the region. A discernible shift from statist economic models to market-oriented policies coincided with a notable rise in democratic governance, forging a nuanced relationship between economic reformation and the growth of liberal democracy. The economic landscape before the 1980s was characterized by Import Substitution Industrialization (ISI), a model promoting domestic industries through protectionist measures such as tariffs and import quotas. However, by the 1980s, many Latin American countries were grappling with external debt, hyperinflation, and stagnant growth (Cardoso and Faletto 1979). The exigencies of the situation, amplified by the pressure from international monetary institutions, especially the International Monetary Fund (IMF) and the World Bank, prompted a paradigm shift. The Washington Consensus emerged as a set of policy recommendations emphasizing fiscal discipline, deregulation, liberalization of trade and investment, and privatization of state enterprises (Williamson 1990). Parallel to these economic reforms, the region also experienced a democratizing wave.

The transition from authoritarian regimes, whether military juntas or civilian autocracies, to more democratic forms of governance was palpable. Countries like Brazil, Chile, and Argentina transitioned from military rule to elected governments, while others, like Mexico, witnessed a gradual opening up of the political space leading to competitive multiparty systems (Schmitter and Karl 1991). The confluence of economic and political changes instigated debates among scholars. Some argue that economic liberalization facilitated the democratization process. The reasoning posits that opening up economies created new entrepreneurial classes independent of the state, strengthening civil society and thus propelling demands for political freedoms (O'Donnell 1973). Moreover, as countries integrated into the global economy, there was increased pressure from international actors to adopt democratic norms (Smith 2005). However, the relationship between market reforms and democracy isn't without its complexities. In some cases, the immediate aftermath of economic liberalization led to increased inequalities and social tensions (Stallings and Peres 2000). Moreover, while countries adopted democratic structures, issues like corruption, weak rule of law, and underdeveloped institutions raised concerns about the depth and quality of this democracy (Mainwaring and Welna 2003).

In conclusion, the economic reformation of Latin American countries in the latter part of the twentieth century is inextricably intertwined with the region's democratic trajectory. While economic liberalization and democratization often moved in tandem, the nuanced interplay between them underscores the multifaceted nature of the region's transformation. As Latin America continues to evolve, understanding the synergies and tensions between its economic and political reforms remains pivotal.

3.3.1 Economic Diplomacy and Major Reforms in Latin American Countries

Economic diplomacy, encompassing a state's international efforts to negotiate economic relationships, trade deals, and investment climates, has been a cardinal tool for Latin American countries in their pursuit of economic reform and modernization. Leveraging diplomatic channels, these nations have sought to integrate their economies into global markets, attract foreign investment, and achieve sustainable growth.

Regional Integration

Regional integration is a key part of Latin America's economic diplomacy. The early 1990s founding of MERCOSUR (Mercado Común del Sur) is a monument. MERCOSUR strengthened intra-regional trade and provided a unified front in negotiations with other global economic blocs by promoting free trade and fluid movement of goods, services, and factors of production between its members (Argentina, Brazil, Paraguay, Uruguay, and later Venezuela).

Bilateral and Multilateral Trade Agreements

Latin American governments have also increased bilateral and multilateral trade agreements beyond the area. Chile signed free trade agreements with the EU, US, and China to diversify its trading portfolio and reduce its economic dependence on one market (Aggarwal and Espach 2004). Mexico's 1994 NAFTA membership signified a shift towards exposing its economy to North American markets and global investors.

Foreign Direct Investment (FDI)

Economic diplomacy in Latin America has also been characterized by efforts to attract foreign direct investment (FDI). Countries such as Brazil and Colombia have engaged in diplomacy to secure FDI in sectors ranging from infrastructure to technology. These efforts are often tied to wider economic reform agendas aimed at improving the investment climate, ensuring legal guarantees for investors, and streamlining bureaucratic procedures (Jenkins 1995).

Engagements with International Financial Institutions

Latin American nations have used economic diplomacy to resolve financial crises and engage with international financial institutions. The region's participation with the IMF and World Bank, especially during debt crises, illustrates the diplomatic challenge of gaining financial aid while maintaining economic and political stability (Kay 2017).

Challenges and Critiques

However, economic diplomacy has proved difficult in the region. Regional trade blocs seek economic integration, yet national interests and economic imbalances can hinder collaboration (Fawcett 2004). Additionally, external trade treaties have sometimes provoked domestic concerns about their effects on local businesses and jobs. Economic diplomacy shaped Latin American economic reforms. This region's states have promoted economic resilience and prosperity through regional integration, trade agreements, investment drives, and global financial institution interactions. The effectiveness of Latin American economic diplomacy will affect their economic outcomes as global economic forces change.

3.4 Bilateral Issues Among Latin American Countries

Bilateral relations in Latin America, as in other regions, are shaped by historical, territorial, economic, and political dynamics. These interactions have sometimes led to disputes, negotiations, and partnerships. A clear understanding requires a deep dive into specific country pairs to unpack the intricacies of their bilateral ties. Bilateral issues among Latin American countries reflect a tapestry of historical legacies, competing national interests, and the imperatives of regional cooperation. Addressing these challenges requires a nuanced understanding of the historical and geopolitical contexts, as well as a commitment to diplomatic resolution mechanisms.

Territorial Disputes

Chile and Bolivia: One of the most protracted territorial disputes in Latin America is between Chile and Bolivia over the latter's access to the Pacific Ocean, a result of the War of the Pacific (1879–1883). Bolivia lost its coastal territory, and despite subsequent treaties like the Treaty of Peace and Friendship in 1904, the quest for sovereign sea access remains a sensitive issue in Bolivian diplomacy.

Economic and Trade Issues

Brazil and Argentina: As the two largest economies in South America, Brazil and Argentina have had both cooperative and competitive economic interactions. While MERCOSUR has facilitated trade, disagreements over tariffs, trade balances, and industrial policies have occasionally strained relations.

Environmental and Water Resources

Paraguay and Brazil: The Itaipu Dam on the Paraná River is the world's second-largest hydroelectric project. While it stands as a symbol of cooperation, disputes over energy pricing and revenue distribution have arisen, given its significance for both nations' electricity needs.

Drug Trafficking and Border Security

Colombia and Venezuela: Geopolitical tensions combined with issues of drug trafficking and illegal border crossings have led to diplomatic strains. Accusations of harboring rebels and not doing enough to curb drug trade have been exchanged, necessitating bilateral dialogues for resolution.

Ideological and Political Differences

Venezuela and Guyana: Historical territorial claims over the Essequibo region have been exacerbated by ideological differences. Venezuela's socialist government and Guyana's westward orientation have occasionally intensified their territorial disagreements.

Cultural and Historical Grievances

Mexico and Guatemala: These nations share cultural ties, given their Mayan heritage. Yet, historical issues such as the undocumented migration of Guatemalans to Mexico and the treatment of these migrants have strained relations. The shared border sees human trafficking, contraband, and has been the center of various security discussions.

Maritime Boundaries and Fisheries

Peru and Chile: Maritime disputes have surfaced between these countries, particularly concerning the rich fishing grounds off their coasts. The International Court of Justice's (ICJ) 2014 ruling tried to settle the boundary, but complexities concerning fisheries management persist.

Energy and Natural Resources

Argentina and Uruguay: The pulp mill disputes over the Rio Uruguay involved concerns about the environmental impacts of a Finnish paper mill on the Uruguayan side. Argentina claimed that the mill could harm the river's biodiversity and negatively affect tourism, leading to a case at the ICJ.

Defense and Military Concerns

Ecuador and Peru: The Cordillera del Cóndor region was a flashpoint due to territorial disagreements dating back to Spanish colonial times. Skirmishes flared as recently as 1995, but diplomatic efforts, particularly the Brasilia Presidential Act, have brought a resolution.

Shared Ecosystem and Conservation

Bolivia and Brazil: The Pantanal, the world's largest tropical wetland area, spans these countries. While it's a UNESCO World Heritage site, deforestation, especially on the Bolivian side, has raised concerns in Brazil, which views the Pantanal as a vital ecological buffer zone.

The intricacies of bilateral relationships in Latin America underscore the importance of diplomacy, dialogue, and international mediation. As these nations navigate the paths of development and regional integration, addressing these bilateral issues becomes paramount for the broader aspirations of peace and cooperation in the region.

3.5 Major Economic Resource and Market Systems of Latin America

Latin America, a region replete with natural resources and varied economic systems, presents a captivating study of economic potential and diversification. The countries, while sharing certain regional traits, boast distinct resource endowments and market mechanisms that influence their global economic standing. Latin America has vast and varied geography, possesses a rich array of economic resources that have historically driven its economies and dictated its interactions on the global stage. Here, we delineate these resources by category and their prominence across different nations.

Minerals and Metals

Brazil: This nation boasts a plethora of minerals, including iron ore, bauxite, gold, and manganese. The Carajás Mine, the world's largest iron ore mine, is located in Brazil.

Chile: Primarily known for its copper, Chile is the world's top producer and exporter of this metal. The country is home to the vast Escondida mine, which produces the highest volume of copper globally (Cochilco 2019).

Bolivia: Bolivia is known for its significant silver, tin, and lithium deposits. The Salar de Uyuni is recognized as the world's largest lithium reserve, vital for the burgeoning electric vehicle industry (Gruber et al. 2011).

Agricultural Produce

Argentina: Globally recognized for its beef, Argentina's Pampas region provides fertile land ideal for cattle ranching. The nation is also a major producer of soybeans, maize, and wine (Richards 2012).

Brazil: Besides being the leading producer of coffee, Brazil is also renowned for its production of sugarcane, primarily for sugar and ethanol, and soybeans (FAO 2018).

Mexico: Mexico is the birthplace of maize, and it remains a central agricultural product. Additionally, the country is a significant producer of tomatoes, avocados, and tequila (FAO 2018).

Energy Resources

Venezuela: Known for its vast oil reserves, the Orinoco Belt in Venezuela holds one of the world's largest accumulations of heavy crude oil (Mommer 2002).

Brazil: Apart from petroleum, Brazil has made significant strides in biofuel production, particularly ethanol derived from sugarcane (Goldemberg 2008).

Mexico: The country has substantial oil reserves, with the Cantrell Field being one of its primary production sites, though its output has been declining in recent years (EIA 2020).

Forest Resources

Brazil and Peru: Both nations, encompassing vast tracts of the Amazon rainforest, possess abundant timber and non-timber resources. However, concerns of deforestation and indigenous rights are pivotal in their management (Barbier and Burgess 2001).

Textiles

Bolivia & Peru: Both are celebrated for their high-quality alpaca and llama wool. These fibers, known for their warmth and softness, command premium prices in international markets.

Colombia: The nation is recognized for its fashion and textile industry, particularly in areas like Medellín.

Fishing and Aquaculture

Chile: Apart from its mineral wealth, Chile has a robust fishing and aquaculture sector, being one of the world's largest producers of fish and seafood, especially salmon (FAO 2020).

Peru: The Humboldt Current makes the Peruvian sea one of the richest in the world, supporting an enormous fishing industry. Anchovy fishing supports Peru's fishmeal and fish oil industry, the largest globally (Fréon et al. 2008).

Manufacturing and Industry

Mexico: Beyond primary commodities, Mexico has a diverse manufacturing sector, especially in automobiles, textiles, and electronics. The North American Free Trade Agreement (NAFTA) significantly influenced its integration into global manufacturing supply chains.

Brazil: The country has a substantial manufacturing base, with industries ranging from automobiles, aircraft (Embraer is one of the largest airplane manufacturers), and durable goods.

Tourism

Mexico, Dominican Republic, and Costa Rica: These countries have prominent tourism sectors, with attractions ranging from historical sites, such as the Mayan ruins, to natural ecosystems like rainforests and beaches. The sector plays a significant role in their economies (Honey 2008).

Renewable Energy

Costa Rica: Noteworthy for its commitment to renewable energy, Costa Rica frequently generates a substantial portion of its electricity from renewable sources, mainly hydroelectric power, but also wind, solar, and geothermal energy.

Brazil: The country is a leader in hydroelectric power, with Itaipu Dam being one of the largest producers of hydroelectric power in the world. Additionally, Brazil's focus on wind and solar power generation is growing.

The economic resources of Latin America are diverse and abundant, ranging from traditional sectors like agriculture and mining to burgeoning areas such as renewable energy and tourism. Each country in the region contributes uniquely to the collective wealth. However, moving forward requires a focus on sustainable practices, environmental conservation, and policies that promote equitable economic benefits for their populations. The economic potential of Latin America, founded upon its rich resource endowment, places the region in a critical position in global supply chains. Yet, judicious management, sustainable utilization, and equitable distribution remain challenges to be addressed.

3.5.1 Market Systems of Latin America

Market systems, commonly understood as the structures and mechanisms that determine how goods and services are bought and sold, have evolved distinctly within Latin American countries. Influenced by their colonial history, post-independence policies, and international market dynamics, these systems have shown a dynamic mix of interventionist and free-market tendencies.

1. **Import Substitution Industrialization (ISI)**

From the 1930s through the 1980s, many Latin American countries adopted ISI as an economic strategy. Here, the focus was on reducing external dependence by promoting domestic industries. This was achieved through protective tariffs, import quotas, and state intervention in strategic sectors. While countries like Brazil, Mexico, and Argentina witnessed industrial growth under ISI, they also accumulated external debt and experienced inefficiencies due to lack of competition (Baer 1972).

2. **Shift to Neoliberal Policies**

In the late twentieth century, particularly during the 1990s, there was a region-wide shift towards neoliberal policies under the "Washington Consensus". These policies advocated for open markets, deregulation, and privatization. Countries like Chile under Pinochet became poster children for these free-market reforms. However, these policies also faced criticism for exacerbating inequalities and undermining domestic industries (Edwards 1995).

3. **Commodity Exporting Model**

Countries like Venezuela (oil), Chile (copper), and Brazil (soybeans and iron ore) have economies significantly influenced by commodity exports. These countries' market systems are highly dependent on global commodity prices. This dependency makes them susceptible to boom and bust cycles, often driven by factors beyond their control (Bulmer-Thomas 2014).

4. **Regional Trade Blocs**

To enhance their bargaining position and reduce external vulnerabilities, Latin American countries have formed regional trade blocs like Mercosur and the Pacific Alliance. These blocs not only aim to boost intra-regional trade but also to harmonize economic policies across member states (Phillips 2014).

5. **Informal Markets**

Many Latin American economies rely heavily on the informal market. Many people in developing nations like Bolivia and Peru make their living through unofficial channels like street vending and unofficial companies. Despite their importance to the economy in terms of employment and output, informal markets are notoriously difficult to regulate and sometimes lack access to official finance or legal safeguards (Tokman 2007).

6. **Contemporary Shifts and Challenges**

Over the past few years, China has become a significant trading partner for numerous Latin American countries, expanding its influence in the region. While this has resulted in investment and growth, it has also raised questions about the sustainability of debt (Jenkins 2016).

7. **Agricultural Export Model**

Latin American countries relied on the export of agricultural commodities like coffee, bananas, and sugar. Some countries, like Guatemala, were extremely dependent on banana exports, while others, like Brazil and Colombia, were extremely dependent on coffee exports (Bates 1981).

8. **Mixed Economies and State Enterprises**

Several nations in Latin America have sizable state-run businesses, such as Mexico's PEMEX (oil) and Brazil's PETROBRAS (oil). These firms exhibit a hybrid of market and state control and tend to dominate economically significant industries (Ramamurti 1999).

9. **Financial Markets**

Latin American countries have witnessed significant changes in their financial sectors. Financial liberalization in the 1990s led to a surge in capital inflows but also to crises, as seen in the Mexican "Tequila Crisis" of 1994. More robust regulatory frameworks have since been established, but challenges persist in terms of financial deepening and inclusion (Montiel 2003).

10. **Service Sector and Tourism**

The service industry, which includes the travel industry, is vital but often gets overlooked in favor of commodities or manufacturing. Millions of tourists go to countries like Costa Rica, the Dominican Republic, and Mexico every year because of the flourishing tourism industry there. The country's GDP and foreign exchange revenues benefit greatly from the flood of visitors (Honey 2008).

11. **Technological Innovation and Start-Ups**

So, Paulo, Buenos Aires, and Mexico City are becoming important centers of technological development in South and Central America. With the help of local and global venture capital, more and more startups and tech enterprises are cropping up, signaling a move towards knowledge-based market segments (Melo and Rodrguez-Pose 2020).

12. **Social Markets and Indigenous Trade**

Traditional and social markets play a significance in countries with sizable indigenous populations like Bolivia, Peru, and Ecuador. Local economies and cultural continuity rely heavily on these markets, which frequently adhere to distinct social norms and conventions. The intricate web of market systems in Latin America reflects the region's cultural diversity, historical depth, and wide range of economic activities. The global and regional influences on these systems mean that they are always changing to meet new demands and take advantage of emerging opportunities. Both domestic policy initiatives and global economic forces have contributed to the development of Latin America's market systems. The region's market systems will continue to develop as it responds to the problems of the twenty-first century, such as technology disruption and environmental concerns.

3.6 International Trade of Latin American Countries: Major Export and Import

The term 'Latin American countries' (LAC) encompasses a wide range of countries, each with its own distinct economy, history, and culture. Using information from trustworthy sources including the World Bank, the International Monetary Fund, and regional economic organizations, this chapter presents a comprehensive economic analysis of selected LAC nations (Table 3.1).

Table 3.1 Major international trade products and partners

Country	Major exports	Major imports	Primary trade partners
Brazil	Soybeans, iron ore, petroleum, chicken meat, sugar, coffee, airplanes	Machinery, chemical products, oil, electronics, automotive parts	China, USA, Argentina, Netherlands, Germany
Mexico	Automobiles, electronics, petroleum products, silver, fruits, vegetables	Machinery, electrical/electronic equipment, steel	USA, Canada, China, Japan, South Korea
Argentina	Soybeans, corn, wheat, petroleum, gold, wine	Machinery, vehicles, plastics, chemicals	Brazil, China, USA, Chile, Vietnam
Chile	Copper, fish products, grapes, chemicals, wine	Machinery, petroleum, electronics, vehicles	China, USA, Japan, South Korea, Brazil
Colombia	Petroleum, coffee, coal, gold, flowers	Machinery, transport equipment, oils, electronics	USA, China, EU, Mexico, Brazil
Peru	Copper, gold, textiles, fishmeal, petroleum	Machinery, petroleum, electronics, plastics	China, USA, Canada, South Korea, Brazil
Venezuela	Petroleum, bauxite, aluminum, steel, chemicals	Food, medicine, machinery, consumer goods	China, India, Russia, USA, Cuba (given political alignments)
Ecuador	Petroleum, bananas, shrimp, coffee, cacao	Vehicles, telecommunications, medicines	USA, China, EU, Chile, Peru

Source Economic Commission for Latin America and the Caribbean (ECLAC)

3.6.1 Major International Trade Blocs and Agreements

International trade is pivotal to the economic growth and sustainability of Latin American nations. Several trade blocs and agreements have been initiated to foster economic integration and trade relations both within the region and with the world. This essay will discuss some of the most significant trade blocs and agreements in Latin America.

1. **Mercosur (Mercado Común del Sur)**

Formation: Established in 1991 through the Treaty of Asunción.

Member Nations: Primarily includes Argentina, Brazil, Paraguay, Uruguay, and Venezuela (though Venezuela's membership was suspended in 2016).

Objective: The bloc aims to facilitate free trade and the fluid movement of goods, people, and currency among member countries.

Significance: Mercosur represents a significant portion of the world's GDP and acts as an influential player in global trade negotiations.

Source: https://treaties.un.org/pages/showDetails.aspx?objid=08000002800a84a2

2. **Pacific Alliance**

Formation: Launched in 2011.

Member Nations: Chile, Colombia, Mexico, and Peru.

Objective: Unlike other Latin American trade agreements, the Pacific Alliance focuses not only on trade but also on deeper economic integration. It seeks the free circulation of goods, services, capital, and people among member countries.

Significance: The bloc emphasizes ties with the Asia–Pacific region, recognizing the rising significance of Asian markets.

3. **CAFTA-DR (Dominican Republic-Central America FTA)**

Formation: Became effective for member countries between 2006 and 2009.

The United States, Costa Rica, El Salvador, Guatemala, Honduras, Nicaragua, and the Dominican Republic are the members.

Goal: By removing trade restrictions and tariffs, the agreement will enable more investment and trade among its member nations. The United States and the Central American region use it as one of their main economic engagement tools.

Source: https://investmentpolicy.unctad.org/international-investment-agreements/treaties/tips/5056/cafta---dr-2004-

4. **ALBA (Bolivarian Alliance for the Peoples of Our America)**

Date of formation: 2004.

Member Nations: Contains nations such as Nicaragua, Bolivia, Cuba, Venezuela, and Bolivia.

Goal: Although ALBA emphasizes economic integration, it also has a strong political orientation, with a focus on left-leaning ideals that integrate social and cultural integration.

Significance: Unlike other blocs, which focus on the free market, ALBA offers an alternative vision of integration.

5. **CELAC (Community of Latin American and Caribbean States)**

Incorporated in 2011. Consists of all countries in the Americas, excluding the United States and Canada. Encourage the political, cultural, and economic integration of the states in the Caribbean and Latin America.

Significance: With little influence from North America, CELAC serves as a regional platform for discussion and collaboration (Table 3.2).

The emergence and development of trade blocs and agreements in Latin America is indicative of the region's aspiration to capitalize on the advantages of economic integration. Despite the fact that these accords have promoted commerce and collaboration, issues like political division, economic inequality, and outside influences still exist. Any discussion of Latin America's economic destiny must start with a thorough understanding of these blocs.

Table 3.2 Macroeconomic indicators (2019–2021) of Latin Nations

Country	GDP (USD Billion)	GDP growth rate (%)	Inflation rate (%)	Unemployment rate (%)
Brazil	$1,448.33	– 3.3	3.2	14.1
Mexico	$1,268.71	– 8.2	3.6	4.9
Argentina	$385.51	– 9.9	36.1	10.1
Colombia	$309.40	– 7.2	1.9	16.8
Chile	$274.91	– 5.8	2.7	11.0
Peru	$222.15	– 11.1	2.3	6.6
Venezuela	$74.08	– 35.0	2746.5	23.7

Source World Bank's World Integrated Trade Solution (WITS) (2022)

3.7 Latin American Countries: Balance of Economic Relations Between Western Blocs and China

Traditionally, Western powers—especially the United States and Europe—have anchored Latin America's place in the world economy. Trade agreements, international aid, and the entwined political interests resulting from the Monroe Doctrine have historically defined this relationship (Smith 2012). Nevertheless, when China significantly increased its commercial connections with the area at the start of the twenty-first century, a new dynamic emerged. China's engagement in Latin America, predominantly driven by its insatiable appetite for raw materials, has been marked by significant trade volumes, investment flows, and infrastructure projects (Jenkins 2012). This Sino-Latin American bond is exemplified by the surge in trade volume from a mere \$12 billion in 2000 to more than \$260 billion by 2019 (Gallagher and Myers 2019). Commodities like soybeans, oil, and iron ore form the bedrock of this trade relationship.

The economic overture of China has implications for Latin America's balance of economic relations. On one hand, it offers countries an alternative to traditional Western economic dependencies, potentially translating to better terms of trade and diversified economic portfolios. On the other, the largely commodities-based relationship with China runs the risk of reinforcing Latin America's role as a primary goods supplier, hindering the region's aspirations of moving up the value chain (Dussel Peters 2019). Western powers, recognizing the strategic importance of Latin America, have responded to China's advances with renewed economic engagement efforts. This includes updated trade agreements, investments in sectors beyond raw commodities, and diplomatic initiatives aimed at reaffirming historic ties (Ray and Gallagher 2020). Latin America's economic dance between China and Western powers is emblematic of its strategic importance in the global order. As the region navigates this multifaceted relationship, striking a balance that ensures sustained growth, diversification, and autonomy remains paramount.

3.7.1 Economic Relations Between the USA and Latin American Countries: A Multi-Faceted Perspective

The United States and Latin America share a history spanning several centuries, marked by both collaboration and contention. This relationship's economic dimensions have evolved significantly over time, encompassing trade, investments, aid, and financial diplomacy.

Trade Relations: Historically, the United States has been one of Latin America's most important trading partners. The North American Free Trade Agreement (NAFTA) with Mexico and the United States, Mexico, and Canada Agreement

(USMCA) were the results of efforts to formalize these relations in the late twentieth and early twenty-first centuries (Villarreal and Fergusson 2017). In addition, the United States has signed Free Trade Agreements (FTAs) with a number of countries, including Chile, Peru, and a number of countries in Central America (CAFTA-DR) (Hornbeck 2012).

Foreign Direct Investment (FDI): U.S. multinational corporations have historically played dominant roles in sectors like mining, agriculture, telecommunications, and manufacturing across Latin America (Stallings 1995). While investments have diversified in recent decades, they still remain significant in sectors like energy, technology, and finance.

Development Aid and Financial Diplomacy: U.S. foreign aid to Latin America, delivered through agencies like USAID, has historically focused on promoting economic development, good governance, and counternarcotics (Meyer and Taft-Morales 2019). Additionally, the U.S. has exercised financial diplomacy through institutions like the International Monetary Fund (IMF) and the World Bank, where it holds significant influence.

Migration and Remittances: Labor migration, particularly to the U.S., has economic implications. Remittances sent by Latin American migrants back to their home countries constitute significant financial inflows, providing essential economic support for many families and contributing to national GDPs (Orozco and Garcia-Zanello 2009).

Challenges and Contentions: Historically, economic relations have not been devoid of tensions. Issues such as the U.S. role in supporting certain regimes, trade imbalances, or concerns over U.S. corporate behavior in the region (e.g., banana companies or mining corporations) have sometimes strained relationships (Schoultz 1998).

U.S.-Latin America economic ties are multi-dimensional, reflecting a mix of mutual interests and inherent challenges. As both regions evolve in the context of a rapidly changing global landscape, the relationship demands continual re-evaluation and recalibration.

3.7.2 Economic Relations Between Europe and Latin American Countries: A Comprehensive Examination

The economic ties between Europe and Latin America have deep roots, tracing back to the era of colonization. These ties have undergone profound transformations over the centuries, with modern relationships characterized by trade, investment, development cooperation, and multilateral dialogues.

Trade Relations: The European Union (EU) is a significant commercial partner for Latin American countries. Improved trade links in commodities, services, and intellectual property rights have resulted from bilateral and regional trade agreements like the EU-MERCOSUR Association Agreement, the EU-Chile Association Agreement, and the EU-Mexico Global Agreement.

Foreign Direct Investment (FDI): European countries, especially Spain, have been prominent investors in Latin America. Major sectors attracting European FDI include telecommunications, banking, energy, and manufacturing. Over the years, European investments have supported infrastructure development, technology transfer, and job creation in the region.

Development Cooperation: The European Commission and EU member states have been significant donors for development initiatives in Latin America. This cooperation has prioritized areas like sustainable development, human rights, climate change, and institutional strengthening. Initiatives like the EU-Latin America Regional Strategy have been pivotal in this regard.

Private Sector Engagement: Besides large multinationals, small and medium-sized enterprises (SMEs) from Europe have shown a growing interest in the Latin American market. Business forums and chambers of commerce have been instrumental in fostering these ties, facilitating partnerships, joint ventures, and technology exchanges.

Cultural and Historical Linkages: Economic relations are deeply intertwined with cultural and historical connections, especially between former colonial powers and their former colonies. Such shared histories have influenced trade preferences, investment destinations, and tourism exchanges.

Challenges and Prospects: Despite the robust economic ties, challenges like regulatory disparities, protectionist tendencies, and geopolitical shifts persist. Nevertheless, shared values concerning democracy, human rights, and sustainable development provide a solid foundation for future economic collaboration.

Europe and Latin America's economic relationship remains dynamic, shaped by historical contexts and contemporary geopolitical and economic realities. While challenges persist, the multi-faceted nature of this relationship suggests ample opportunities for further mutual growth.

3.7.3 China's Emerging Trade Growth in Latin American Countries

In the landscape of global economic realignments, China's burgeoning trade relationship with Latin American countries stands out as a remarkable evolution. Beginning the twenty-first century with relatively modest economic ties, the ensuing decades have witnessed an explosive growth in Sino-Latin American trade, underscoring a significant shift in the region's economic dynamics.

The fast industrialization and urbanization that characterized China's growth trajectory created an insatiable need for raw materials, a demand that Latin America's abundance of natural resources was well-suited to meet (Jenkins 2012). China's commerce with Latin America has grown from roughly $10 billion in 2000 to over $300 billion in 2019 (Dussel Peters 2020) as commodities such as soybeans from Brazil and Argentina, copper from Chile, oil from Venezuela, and iron ore from Brazil have become mainstays in the market. China's strategic actions spurred this expansion of international trade. Initiated in 2014, the Forum of China and the Community of Latin American and Caribbean States (CELAC) established a meeting ground for international negotiations. Furthermore, the Belt and Road Initiative (BRI), which was started in 2013, enticed Latin American countries by promising to improve their infrastructure, provide investment possibilities, and strengthen commercial relations (Myers and Wise 2019). However, the structure of this trade growth has raised concerns. While Latin America mainly exports primary goods to China, it imports manufactured goods, leading to concerns about reinforcing a traditional, dependent development model. Such an arrangement can perpetuate Latin America's position as a supplier of raw materials, possibly stymying its ambitions of industrial diversification (Gallagher 2016). Simultaneously, China's competitive manufactured products have posed challenges for Latin America's domestic industries. There's evidence suggesting that increased Chinese competition adversely impacted the region's manufacturing sector (Mesquita Moreira 2013). China's surge in trade with Latin America epitomizes its global economic ascent and the consequential recalibration of international economic relations. While the growth brings opportunities for Latin America in terms of market access and investments, the challenges it presents in terms of development models and domestic industry competition cannot be overlooked. As this relationship matures, it will be crucial for Latin American nations to negotiate terms that ensure sustainable and diversified growth.

3.7.4 USA Concerns Regarding China's Growing Trade Influence in Latin America

The growing economic connection between China and Latin America, exemplified by the increasing trade links and geopolitical alliances, has captured the attention of policymakers in Washington. Throughout history, the Western Hemisphere has been widely regarded as the sphere of influence for the United States, as established by the Monroe Doctrine. The rise of China in this region has consequently raised significant geopolitical and economic apprehensions.

Strategic Concerns: China's increasing footprint in Latin America, a region geographically proximate to the U.S., poses potential security challenges. U.S. policymakers fear that China's influence could challenge the region's alignment with American values and interests, particularly if nations prioritize Beijing's preferences over Washington's.

Economic Concerns: America has been Latin America's biggest trading partner. Brazil, Chile, and other regional countries now trade more with China than the U.S. Such adjustments could restrict market access and regional economic policy impact for U.S. enterprises and the economy.

Ideological and Governance Implications: The United States is worried that China's "no strings attached" approach to trade and investment, which frequently eschews criteria related to governance and human rights, could undermine American initiatives to advance democratic government, openness, and human rights in the area.

Resource Accessibility: There are worries that China's voracious demand for goods from Latin America may result in a shortage or higher price of vital resources for American industries, including rare earth minerals or agricultural products.

Multilateral Dynamics: Regional organizations such as the Organization of American States (OAS) are crucial to the diplomatic framework of the Western Hemisphere, according to the United States. China's establishment of alternative multilateral platforms, such as CELAC, raises concern about the dilution of U.S. influence within traditional regional diplomatic structures.

In response to these concerns, the United States has taken steps to reaffirm its commitment to the region, as exemplified by programs such as America Crece, which aims to encourage private sector investment in energy and infrastructure in Latin America. As China continues its economic expansion into Latin America, the United States confronts the challenge of recalibrating its strategy to preserve its historical ties and influence in the region.

3.8 Economic Diplomacy of Latin America: Case Studies of Brazil, Argentina

Latin America's economic diplomacy is characterized by a rich tapestry of regional alliances, bilateral trade agreements, and strategic engagements with global economic actors. This dynamic interplay is influenced by historical contexts, socio-political landscapes, and a quest for enhanced growth and integration.

Regional Alliances—Mercosur and Pacific Alliance: With its founding in 1991, Mercosur has been at the forefront of economic diplomacy in the region, particularly among its core members of Brazil, Argentina, Uruguay, and Paraguay (Malamud 2005). Together, Chile, Colombia, Mexico, and Peru form the Pacific Alliance, a market-oriented coalition that promotes greater economic integration and involvement with Asia and the Pacific.

Bilateral and Global Engagements: Latin American nations have actively fostered relationships with major economies such as China, the U.S., and the European Union. China's surging investments in the region, especially in infrastructure and commodities, signify a critical shift in the economic diplomacy matrix (Jenkins 2012).

South-South Cooperation: In order to promote a more equitable global economic system and diversify dependencies, nations like Brazil and Argentina have pushed for increased economic and political cooperation within the Global South.

Multilateral Forums: Latin American countries have been vocal in international organizations like the World Trade Organization (WTO), where they have pushed for more equitable trade policy and development concerns. When they bargain as a group, they highlight the importance of fairness in international economic rules.

In conclusion, the economic diplomacy of Latin America displays a mix of regional solidarity, purposeful global outreach, and a strong voice in multilateral economic settings, driven by a desire for sustainable prosperity and equitable global economic systems.

3.8.1 Brazil's Economic Diplomacy: An Overview

As the largest economy in South America, Brazil has employed strategic economic diplomacy to leverage its global influence and achieve its national development goals. This diplomacy is characterized by regional integration, engagement with global powers, promotion of South-South relations, and a proactive role in multilateral economic institutions.

Regional Integration: Brazil has been an integral part of Mercosur (the Southern Common Market) since its creation in 1991. The group's stated goal is to ease the flow of goods, people, and cash among its member states through unrestricted commerce and travel. Mercosur has been a useful tool for Brazil in its efforts to negotiate and solidify regional leadership through large-scale trade agreements.

Engagement with Global Powers: Brazil has actively sought to diversify its trade partners, engaging deeply with countries like China, the U.S., and the European Union. Such relations not only boost trade but also attract foreign investments essential for infrastructure and development projects (Jenkins 2012).

South-South Cooperation: Brazil has become an outspoken advocate for closer economic and political relations among developing countries. Brazil is committed to this policy, as evidenced by its participation in forums such as IBSA (India, Brazil, and South Africa) and BRICS (Brazil, Russia, India, China, and South Africa), which aim to challenge the existing Western-dominated global economic order.

Multilateral Economic Institutions: The World Trade organization (WTO) and the United Nations Conference on Trade and Development (UNCTAD) are two organizations where Brazil has taken the initiative. Its position typically centers on advocating for more equal trade policies and reorganizing international financial systems.

Sustainable Development and Environmental Diplomacy: Given Brazil's vast Amazon rainforest, the country has been at the center of global environmental discussions. Brazil often blends its environmental stance with economic objectives, seeking investments and partnerships that promote sustainable development.

In conclusion, Brazil's economic diplomacy is a blend of strategic regionalism, global engagement, advocacy for the Global South, and leadership in multilateral economic forums. Its approach aims at enhancing its global economic stature while addressing domestic development needs.

3.8.2 Economic Diplomacy of Argentina

Argentina's economic diplomacy reflects its multifaceted engagements at regional, bilateral, and multilateral levels. It encompasses strategies to protect national interests while fostering trade, investment, and international cooperation.

Regional Integration—Mercosur: Argentina has been instrumental in the Mercosur bloc. This integration not only aids in promoting free trade among member nations but also serves as a collective platform to negotiate with larger economies and regions (Malamud and Gardini 2012).

Bilateral Engagements with Major Economies: Argentina has pursued strategic economic ties with global powers. Particularly noteworthy is its evolving relationship with China. From agriculture to infrastructure projects, Chinese investments in Argentina have surged, signifying a shift towards diversifying economic dependencies (Duke 2018).

Debt Diplomacy: Economic diplomacy for Argentina has often revolved around its external debt. Post the 2001 financial crisis, Argentina adopted a stance of renegotiating its debt terms, exemplifying its emphasis on economic sovereignty and strategic engagements with international financial institutions and creditors (Bridges 2006).

Diversified Trade Partnerships: Beyond its traditional partners in Europe and the Americas, Argentina has sought to diversify its trade partnerships, turning towards Asia–Pacific nations and Middle Eastern economies, broadening its economic horizons (Bouzas and Soltz 2001a, b).

Multilateral Forums: Argentina has been proactive in platforms like the WTO and G20, advocating for fairer global economic policies and voicing concerns of emerging economies (Narlikar and Tussie 2004).

In essence, Argentina's economic diplomacy intertwines its domestic economic imperatives with global engagements, underpinned by strategic alliances, diversified trade relations, and a fervent quest for economic autonomy.

Geopolitics of Latin America: Domestic Polarization and Geopolitical Opportunities

In terms of geopolitics and economy, LAC is characterized by a precarious equilibrium between the US's historical domination and China's increasing power. This dynamic presents opportunity as well as a barrier to the EU's increased regional involvement, as it is clear that LAC foreign policy is a complicated patchwork of political autonomy, strategic alignment, and tactical adaptability. The UNGA's voting habits provide a lens through which to view how these nations deftly strike a balance between their national and regional objectives and outside forces. This intricate view serves as a reminder that while attempting to determine the parameters of alignment with the major powers, we cannot entirely rely on economic trends. Both within and outside of them, there are additional factors at work. Therefore, it is critical to keep in mind that LAC's relationship with the big powers is shaped by its relative autonomy and strategic power play, which also provide the way for new collaborations and opportunities.

There is more to the geopolitical and geoeconomic dynamics in the region than just the dichotomy of US versus Chinese dominance. As demonstrated, the range is far more complicated and varies significantly between nations. Every state has varied degrees of political, economic, and historical inclination towards one side or even tries to strike a balance between the two. Because of this, the US has always dominated the region, influencing events through trade agreements, economic assistance, diplomacy, and, occasionally, outright involvement. But at different points in time and different places, this influence has faced criticism and rejection, particularly from countries that have aspired to challenge US hegemony and want greater autonomy. In recent times, China has become an essential economic and political ally for numerous Latin American and Caribbean nations. It has mostly concentrated on economic issues, including commerce, loans, and investments. But China's increasing influence in the region has also sparked worries, especially in relation to matters like investment transparency, environmental sustainability, and economic reliance. Data from the German Institute of Area and Global Studies.

For observers, the most notable feature is the shift in the nature of the relationship between China and Latin America. While China's actions were first motivated by trade opportunities or one-time financial operations, starting in 2013, official policy became the driving force behind them. The "Belt and Road Initiative," an ambitious plan initiated by President Xi Jinping to boost investment and infrastructure projects worldwide, has added Latin America to its list of eligible regions.

References

Aggarwal VK, Espach R (2004) Bargaining over BITs, arbitrating awards: the regime for protection and promotion of international investment. World Politics 56(3):410–441

Baer W (1972) Import substitution and industrialization in Latin America: experiences and interpretations. Lat Am Res Rev 7(1):95–122

Barbier EB, Burgess JC (2001) The economics of tropical deforestation. J Econ Surv 15(3):413–433
Bates RH (1981) Markets and states in Tropical Africa: the political basis of agricultural policies. University of California Press
Bulmer-Thomas V (2014) The economic history of Latin America since Independence. Cambridge University Press
Bouzas R, Soltz H (2001a) Argentina's trade policy in the 1990s: wavering between the WTO and Mercosur. In: Bouzas R, Ros J (eds) Trade, investment, and development in the Americas. Institute of Latin American Studies, University of Miami
Bouzas R, Soltz H (2001b) MERCOSUR: nested, overlapping regionalism. J Interamerican Stud World Affairs 43(1):95–122
Bridges G (2006) The 2001 financial crisis and the restructuring of Argentina's public debt. Dev Chang 37(5):1027–1046
Bushnell D (1993) The making of modern Colombia: a nation in spite of itself. University of California Press
Cardoso FH, Faletto E (1979) Dependency and development in Latin America. University of California Press
Cochilco (Chilean Copper Commission) (2019) Chilean copper mining. Government of Chile, Santiago
Collier S, Sater WF (2004) A history of chile: 1808–2002. Cambridge University Press
Duke, S. (2018). Chinese FDI in Argentina: a closer look at the energy sector. Dialogue of Civilizations Research Institute
Dussel Peters E (2019) China's evolving role in Latin America: can it be a win-win? Atlantic Council Report
Dussel Peters E (2020) China and Latin America and the Caribbean. 2019 Statistical Yearbook. University of Mexico, Center for China-Mexico Studies
Edwards S (1995) Crisis and reform in Latin America: from despair to hope. Oxford University Press
EIA (U.S. Energy Information Administration) (2020) Mexico's Cantarell Field. EIA, Washington, D.C.
FAO (Food and Agriculture Organization) (2018) FAOSTAT statistical database. FAO, Rome
Fawcett L (2004) Exploring regional domains: a comparative history of regionalism. Int Aff 80(3):429–446
Gallagher K (2016) The China Triangle: Latin America's China boom and the fate of the Washington consensus. Oxford University Press
Gallagher KP, Myers M (2019) China-Latin America finance database. Global Trends and Future Prospects, The Dialogue
Gerlach A (2003) Indians, oil, and politics: a recent history of ecuador. SR Books
Goldemberg J (2008) The Brazilian biofuels industry. Biotechnol Biofuels 1(1):6
Gruber G, Medina C, Keolian G, Kesler S, Everson M, Wallington T (2011) Global lithium availability. J Ind Ecol 15(5):760–775
Honey M (2008) Ecotourism and sustainable development: who owns paradise? Island Press
Hornbeck JF (2012) U.S. trade policy and free trade agreements. Nova Science Publishers
Jenkins R (1995) Latin America and Caribbean foreign investment. Third World Quart 16(4):675–698
Jenkins R (2012) Latin America and China—a new dependency? Third World Quart 33(7):1337–1358
Jenkins R (2016) How China is reshaping the global economy: development impacts in Africa and Latin America. Oxford University Press
Kay C (2017) Latin America's failed development and the role of the USA: a new historical perspective. Third World Quart 38(9):2053–2070
Klaren PF (2000) Peru: society and nationhood in the andes. Oxford University Press
Klein HS (2003) A concise history of Bolivia. Cambridge University Press
Lewis PH (1982) Paraguay under stroessner. University of North Carolina Press

Mainwaring S, Welna C (2003) Democratic accountability in Latin America. Oxford University Press

Malamud A, Gardini GL (2012) Has regionalism peaked? The Latin American quagmire and its lessons. Int Spectator 47(1):116–133

Melo PC, Rodríguez-Pose A (2020) The geography of innovation in Latin America: from clusters to systems. J Econ Geogr 20(2):485–513

Mesquita Moreira M (2013) Is there a future for manufacturing in Latin America? In Too far to export: domestic transport costs and regional export disparities in Latin America and the Caribbean. Inter-American Development Bank, pp 45–66

Meyer MC, Sherman WL (2007) The course of mexican history. Oxford University Press

Meyer PJ, Taft-Morales M (2019) U.S. Foreign assistance to Latin America and the Caribbean: FY2018 appropriations. Congressional Research Service

Myers M, Wise C (2019) The political economy of China-Latin America relations in the new millennium. Routledge

Mommer B (2002) Global oil and the nation state. Oxford University Press, Oxford

Montiel PJ (2003) Macroeconomics in emerging markets. Cambridge University Press

Narlikar A, Tussie D (2004) The G20 at the Cancun ministerial: developing countries and their evolving coalitions in the WTO. The World Economy 27(7):947–966

O'Donnell G (1973) Modernization and bureaucratic-authoritarianism. Institute of International Studies

Orozco M, Garcia-Zanello E (2009) Hometown Associations: transnationalism, philanthropy, and development. Brown J World Affairs 15(2):57–73

Phillips N (2014) the southern cone model: The political economy of regional capitalist development in Latin America. Routledge, London

Ramamurti R (1999) Why haven't developing countries privatized deeper and faster? World Dev 27(1):137–155

Ray R, Gallagher KP (2020) China and the Inter-American Development Bank: from borrower to lender. China Q 244:940–959

Richards P (2012) Race and the Chilean miracle: neoliberalism, democracy, and indigenous rights. University of Pittsburgh Press, Pittsburgh

Rock D (1987) Argentina: 1516–1987. University of California Press

Schmitter PC, Karl TL (1991) What democracy is... and is not. J Democracy 2(3):75–88

Schoultz L (1998) Beneath the United States: a history of U.S. Policy toward Latin America. Harvard University Press

Skidmore TE (1999) Brazil: five centuries of change. Oxford University Press

Smith PH (2005) Democracy in Latin America: political change in comparative perspective. Oxford University Press

Smith PH (2012) Talons of the eagle: Dynamics of U.S.-Latin American Relations. Oxford University Press

Stallings B (1995) Global change, regional response: the new international context of development. Cambridge University Press

Stallings B, Peres W (2000) Growth, employment, and equity: The impact of the economic reforms in Latin America and the Caribbean. Brookings Institution Press

Tokman VE (2007) Modernizing the informal sector. UN Chron 44(2):27

Villarreal MA, Fergusson IF (2017) The North American free trade agreement (NAFTA). Congressional Research Service

Weinstein M (1975) Uruguay: democracy at the crossroads. Westview Press

Williamson J (1990) Latin American adjustment: how much has happened? Institute for International Economics

Chapter 4
Contemporary and Prospective Economic Conditions of Bangladesh

Abstract The chapter "Contemporary and Prospective Economic Conditions of Bangladesh" offers a sophisticated analysis of the nation's transformation from a fragile post-independence economy into a global exemplar of resilience and progress. It traces Bangladesh's economic journey, emphasizing its robust GDP growth averaging 6.4% annually from 2016 to 2021, and its projected ascent to a high-middle-income country by 2031. Central to this trajectory is the ready-made garment (RMG) sector, which accounts for over 80% of export earnings, coupled with significant remittance inflows exceeding $18 billion annually. Strategic infrastructure investments, like the Padma Bridge and Rooppur Nuclear Power Plant, and burgeoning sectors such as IT and pharmaceuticals, are positioned as pivotal growth drivers. Despite these advancements, the chapter identifies systemic vulnerabilities, such as over-reliance on the RMG sector, political instability, and susceptibility to climate risks. Bangladesh's monolithic export base leaves it exposed to global trade dynamics, while its geographic location amplifies climate-related threats, including rising sea levels and recurrent natural disasters. However, diversification efforts targeting leather, pharmaceuticals, and IT industries are promising alternatives, aimed at mitigating these risks and fostering economic resilience. A detailed exploration of Bangladesh's trade composition underscores its increasing global integration. Exports of RMG, jute, leather goods, and pharmaceuticals dominate, while imports focus on raw cotton, machinery, and energy. The data reveals persistent trade deficits, with imports outpacing exports, highlighting the need for enhanced export diversification and competitiveness. Key bilateral partners, including China, India, and the EU, play strategic roles in shaping Bangladesh's economic landscape. Notably, China's Belt and Road Initiative (BRI) and India's historical ties underscore geopolitical and economic dependencies. The chapter also highlights Bangladesh's growing prominence in multilateral economic forums such as SAARC, WTO, and BIMSTEC, where it advocates for equitable trade policies and regional connectivity. Additionally, mega development projects and Export Processing Zones (EPZs) are discussed as transformative initiatives enhancing infrastructure, generating employment, and attracting foreign direct investment. The Padma Bridge alone is estimated to contribute an additional 1% to GDP, showcasing the economic ripple effects of such ventures. Bangladesh's economic reforms are critically analyzed within the context

M. T. Islam and M. B. Hossen, *Economic Diplomacy*, The Latin American Studies Book Series, https://doi.org/10.1007/978-3-031-81362-7_4

of its political economy. From post-independence nationalization to market liberalization in the 1980s, the chapter highlights policy adaptations that reflect shifting domestic and global imperatives. Challenges in governance, infrastructure, and financial sector stability remain, necessitating strategic policy interventions. Bangladesh's economic narrative exemplifies a dynamic interplay of challenges and opportunities. As the nation aspires for sustainable and inclusive growth, strategic diversification, innovation, and resilience-building will be essential to navigating global economic shifts and achieving its development goals. This analysis provides a comprehensive framework for understanding Bangladesh's evolving economic paradigm within a globalized context.

Keywords RMG · Market Liberalization · Economic Forums · GDP growth

4.1 Introduction

Bangladesh, once termed a "basket case" by Henry Kissinger, has now morphed into an exemplary testament of economic progress and resilience (Alamgir 1972). The economic ascension of the country can be attributed to various strategic initiatives, paramount among which is the rapid expansion of the ready-made garment (RMG) sector. However, like all developing economies, the country faces both intrinsic challenges and immense opportunities. This brief analysis offers an overview of the current economic landscape of Bangladesh and speculates on prospective economic trends.

Bangladesh's average annual GDP growth of 6.4% between 2016 and 2021 was significantly higher than the global average of 2.9% and far ahead of other Asian countries like India, Indonesia, and the Philippines. The country's per capita GDP of $2800 in 2022 was higher than India's $3200. If the economy continues to expand at its current rate, Bangladesh will join the ranks of the world's high-middle-income countries by 2031 (with a per capita GDP of at least $4000. Bangladesh has proven to be able to weather economic storms on a global scale. One of the few Asian economies to flourish in 2020 despite the COVID-19 pandemic was Bangladesh, where GDP grew by 3.4%. The annualized growth rate for the country was 5.5% throughout the global financial crisis of 2007–2009, which was significantly higher than the global average of 0.4%. The rapidly growing middle and upper classes in Bangladesh are helping to fuel the country's booming domestic consumer sector, which now accounts for about 70% of GDP. More than 34 million Bangladeshis, or about 15 percent of the population, may fall into this category of affluent customers by 2030. Bangladesh's annual GDP growth rate has been astoundingly stable. With an expected growth rate of around 8% in 2019, it will be one of the world's fastest-growing economies in 2019. The RMG industry has been the primary driver of this expansion. This industry is responsible for over 80% of the country's export revenues.

Remittances have been a significant source of foreign exchange reserves, with Bangladeshi expatriates sending home over $18 billion in 2019. This inflow has provided the essential balance-of-payment support and bolstered domestic consumption. The government's prioritization of massive infrastructure projects, such as the Padma Multipurpose Bridge, aims to enhance connectivity and spur economic activity (Ahmed and Hossain 2018). Simultaneously, investments in energy, including the Rooppur Nuclear Power Plant, signify steps towards addressing the country's burgeoning energy needs. The heavy dependence on the RMG sector presents a monolithic economic structure susceptible to external shocks (Zaman and Akhter 2018). Global economic downturns or shifts in global trade dynamics could disproportionately affect the Bangladeshi economy. Bangladesh is on the front lines of climate change, with rising sea levels and recurrent natural disasters threatening its economic infrastructure and agricultural productivity.

The e-Governance initiative envisions comprehensive digitization of services and infrastructural development in IT. The emergence of technology-driven sectors like IT outsourcing and the e-commerce industry hint at a promising digital future. Efforts are underway to explore and bolster alternative sectors like leather, pharmaceuticals, and agro-based industries. Such diversification can provide a buffer against potential setbacks in the RMG sector (Ministry of Commerce 2020). Bangladesh's strategic location between South and Southeast Asia positions it as a potential connectivity hub. With China's Belt and Road Initiative and the Bangladesh-China-India-Myanmar (BCIM) corridor, there's scope for greater regional integration, promoting transboundary trade and investment (Chowdhury 2017). Bangladesh's economic narrative showcases an undeterred commitment to progress. While contemporary accolades validate its growth trajectory, future success necessitates strategic diversification, technological adaptation, and adept management of ecological challenges.

4.2 The Struggling Pathway of Bangladesh's Economic Progress

Since its independence in 1971, Bangladesh has faced a myriad of challenges on its path to economic progress. Over the decades, this South Asian nation has exhibited resilience, making notable strides, but not without encountering significant obstacles. Bangladesh emerged from a tumultuous liberation war, which left its infrastructure decimated and its population traumatized. The initial year's post-independence was marred by political instability, famine, and economic hardships (Osmani 2004).

Structural Issues: Agriculture has historically been a major contributor to the country's economy. However, cyclones and floods are two of the most common natural calamities that have a lasting effect on crop production. As a result of this reliance, Bangladesh was vulnerable to international price swings and climatic abnormalities (Paul and Rashid 2018).

Industrialization and the RMG Sector: Industrialization increased in the late 1970s and 1980s, particularly in Ready-Made Garments (RMG). By the 1990s, RMG exports drove the economy. However, labor unrest, poor working conditions, and tragedies like the 2013 Rana Plaza collapse have emphasized the need for stronger regulation and worldwide ethical manufacturing standards (Berik and Rodgers 2008).

Economic Reforms: Economic liberalization policies implemented in the 1990s brought both a rise in FDI and a rise in remittances sent home by Bangladeshi workers. Rapidly growing remittances boosted GDP, bolstered foreign exchange reserves, and protected the economy from downturns (Siddiqui and Abrar 2003) (Table 4.1).

Digitalization and Diversification: Technology integration across multiple industries has been a focus of the "e-Governance" campaign in recent years. Despite its youth, the Information Technology (IT) sector is shown potential as a new engine of economic expansion.

Challenges Ahead: Despite these accomplishments, Bangladesh still struggles with systemic problems. These include political instability, corruption, inadequate infrastructure, and a growing population, all of which place a strain on the country's resources (Transparency International 2019).

The economic voyage of Bangladesh is evidence of its resilience and adaptability. From its agricultural beginnings to its global prominence in the RMG sector and its burgeoning digital frontiers, the country's development demonstrates its potential, despite its many obstacles. Managing systemic obstacles and capitalizing on its strengths will be crucial for Bangladesh's economic growth in future.

Table 4.1 Growth performance in the five-year plans and PRSP

Specific plan	Plan period (FY)	Average GDP growth rate (%)	
		Target (%)	Actual (%)
First five-year plan	1973–1978	5.5	4.0
Two-year plan	1978–1980	5.6	3.5
Second plan	1980–1985	5.4	3.8
Third plan	1985–1990	5.4	3.8
Fourth plan	1990–1995	5.0	4.2
Fifth plan	1997–2002	7.0	5.1
Sixth plan	2011–2015	7.3	6.3
Seven Plan	2016–2020	7.4	7.13
Eight Plan	2020–2025	8.51	

Source Bangladesh Bureau of Statistics

4.3 Contemporary Conditions of the Bangladesh Economy

Bangladesh's economy has undergone significant transformation over the past few decades, evolving from an agrarian-based structure to one that has a thriving industrial sector, particularly in ready-made garments (RMG). By the 2020s, several key developments and challenges emerged:

Robust Growth Rates: Bangladesh has been among the fastest-growing economies in the world. In 2019, its GDP grew by approximately 8.2%, propelled by exports and domestic consumption.

Dominance of the RMG Sector: The RMG sector remains the country's most significant foreign exchange earner, contributing to over 80% of its total exports. However, global shifts, including the US-China trade tensions and the aftermath of the COVID-19 pandemic, have posed challenges to this sector (Table 4.2).

Rise in the IT and Telecommunications Sector: With the aim of making strong e-Government, there's been a consistent effort to promote the IT sector. As a result, IT exports and the local software market have seen substantial growth, with IT-enabled services expanding rapidly.

Infrastructure Developments: Bangladesh has prioritized large infrastructure projects like the Padma Multipurpose Bridge, aiming to improve connectivity and spur economic activities.

Shift to Middle-Income Status: Bangladesh has successfully graduated from the World Bank's 'low-income' to 'lower middle-income' category, showcasing its economic advancement.

Challenges in Banking Sector: Non-performing loans and financial irregularities remain major concerns for the banking sector, potentially threatening financial stability.

Environmental and Climate Concerns: Given its geographical location, Bangladesh remains vulnerable to climate-related risks. These risks, from rising sea levels to cyclones, pose considerable challenges to the agricultural sector and urban habitats (Table 4.3).

Demographic Dividend and Labor Market: A significant portion of the population is youthful, offering a potential demographic dividend. However, generating adequate quality employment remains a challenge, emphasizing the need for skills development and education reform.

Bangladesh's economy, in the contemporary period, showcases a blend of notable achievements and inherent challenges. By addressing structural issues and capitalizing on its strengths, Bangladesh is poised to maintain its growth trajectory in the coming years.

Table 4.2 Imports of Bangladesh: at a glance

Items	2017–18	2018–19	2019–20	2020–21	2021–22
Total imports (Million Tk.)	5,511,644	5,722,675	5,441,658	6,633,052	8,939,518
Total imports (Million US $)	67,133	68,103	64,186	78,211	103,586
Growth rate of imports (%)	16.96	3.83	− 4.91	21.89	34.77
Imports as percent of GDP	24.49	22.51	19.46	23.74	22.57
Imports as percent of revenue receipts	212.43	180.75	156.34	188.69	229.81
Per capita imports (Tk.)	33,679	34,568	32,476	39,179	52,186
Imports of consumer goods as (%) of	5.11	2.98	2.58	2.69	11.28
Imports of major commodities (Million TK.)					
Consumer goods	888,037	568,881	540,221	690,414	1,008,356
Materials for consumer goods	3,716,874	4,235,415	4,160,831	5,177,557	7,039,390
Capital goods	844,755	870,972	691,419	707,580	830,670
Materials for capital goods	61,978	47,407	49,187	57,501	61,101
Total	5,511,644	5,722,675	5,441,658	6,633,052	8,939,518
Composition of major imported commodities (%)					
Consumer goods	16.11	9.94	9.93	10.41	11.28
Materials for consumer goods	67.44	74.01	76.46	78.06	78.74
Capital goods	15.33	15.22	12.71	10.66	9.29
Materials for capital goods	1.12	0.83	0.90	0.87	0.68
Total	100.00	100.00	100.00	100.00	100.00
Imports by accounts (Million Tk.)					
Private	4,976,611	5,210,822	4,802,236	5,795,599	7,981,890
Government	32,413	6610	23,357	30,895	27,927
Semi-government	502,620	505,243	616,065	806,558	865,633
TCB	–	–	–	–	64,068
Total	5,511,644	5,722,675	5,441,658	6,633,052	8,939,518
Imports by routes (Million Tk.)					
Sea	5,198,361	5,404,720	5,128,752	6,137,846	8,212,117
Land	313,283	317,955	312,906	495,206	727,401
Total	5,511,644	5,722,675	5,441,658	6,633,052	8,939,518

Source Bangladesh Foreign Trade Statistics 2021–22

4.4 Major Economic Resources of Bangladesh

Bangladesh's economic resources paint a picture of diversity and potential. Leveraging these resources efficiently and sustainably is pivotal for the nation's continued growth and development.

Table 4.3 Exports of Bangladesh: at a glance

Items	2017–18	2018–19	2019–20	2020–21	2021–22
Total exports (Million Tk.)	3,087,936	3,481,620	2,811,668	3,479,890	4,571,354
Total exports (Million US $)	37,612	41,433	33,164	41,032	52,970
Growth rate of exports (%)	2.80	12.75	– 19.24	23.77	31.36
Exports as percent of GDP	13.72	13.69	10.05	12.46	11.51
Per capita exports (Tk.)	18,869	21,031	16,780	20,555	26,686
Export of major commodities (Million Tk.)					
Consumer goods	2,799,366	3,165,488	2,524,550	3,118,774	4,121,727
Materials for consumer goods	238,081	269,654	208,676	298,894	365,975
Capital goods	48,400	40,574	74,927	60,308	81,858
Materials for capital goods	2089	5904	3515	1914	1793
Total	3,087,936	3,481,620	2,811,668	3,479,890	4,571,354
Composition of major exported commodities (%)					
Consumer goods	90.65	90.92	89.79	89.62	90.16
Materials for consumer goods	7.71	7.75	7.42	8.59	8.01
Capital goods	1.57	1.16	2.66	1.73	1.79
Materials for capital goods	0.07	0.17	0.13	0.06	0.04
Total	100.00	100.00	100.00	100.00	100.00
Exports by accounts (Million Tk.)					
Private	3,071,063	3,450,102	2,798,801	3,474,396	4,568,290
Government	36	78	399	496	463
Semi-government	16,837	31,440	12,468	4998	2601
Total	3,087,936	3,481,620	2,811,668	3,479,890	4,571,354
Exports by routes (Million Tk.)					
Sea	3,055,231	3,446,718	2,766,288	3,439,963	4,526,572
Land	32,705	34,901	45,380	39,927	44,782
Total	3,087,936	3,481,619	2,811,668	3,479,890	4,571,354

Source Bangladesh Foreign Trade Statistics 2021–22

Agriculture and Land Resources: Bangladesh, with its fertile deltaic plains, is principally an agrarian economy. Rice is the staple crop, followed by jute, often referred to as the "Golden Fiber" of Bangladesh. The nation was once the world's largest producer of jute. Additionally, wheat, tea, and sugarcane are also significant crops. Aquaculture has flourished due to the country's abundant water bodies, making Bangladesh one of the top fish producers globally.

Human Capital: With over 160 million inhabitants, Bangladesh's primary economic resource is its people. The abundant labor force, particularly in the younger demographic, has been a boon for industries such as ready-made garments (RMG), construction, and services. The remittances sent by Bangladeshi expatriates, working predominantly in the Middle East, have also become a major economic pillar.

Ready-Made Garments (RMG) and Textiles: The RMG sector is the lifeline of Bangladesh's industrial economy, accounting for over 80% of the country's total exports. This dominance stems from the country's competitive labor costs and international trade agreements.

Natural Gas: Bangladesh has significant natural gas reserves, primarily situated in the eastern part of the country. These reserves, apart from domestic consumption, offer potential export opportunities, though they are primarily used for power generation and industries within the country.

Leather and Leather Goods: Capitalizing on the local cattle population, Bangladesh has emerged as a significant exporter of leather and leather goods. The leather industry has grown exponentially, becoming the second-largest export sector after RMG.

Information Technology (IT) and IT-Enabled Services: The government's e-government related initiatives have spurred growth in the IT sector. Software development, business process outsourcing, and IT-enabled services are rapidly growing segments, gaining both local and international traction.

Tourism: While not a major resource yet, the Sundarbans (the world's largest mangrove forest) and the sandy beaches of Cox's Bazar are gaining recognition. The government aims to harness this potential and develop tourism as a significant revenue source.

Handicrafts and Cottage Industries: Traditional handicrafts, such as nakshi kantha (embroidered quilts), pottery, and muslin, have long been integral to Bangladesh's cultural heritage. Many rural communities rely on these crafts for their livelihood, and they've gained international recognition for their uniqueness and quality (Rahman 2017).

Maritime Resources: Bangladesh has an extensive Exclusive Economic Zone (EEZ) and recently won rights over a vast maritime area in the Bay of Bengal. This offers prospects for deep-sea fishing, exploration of marine minerals, and the establishment of maritime industries.

Pharmaceuticals: The pharmaceutical sector in Bangladesh has witnessed impressive growth. By producing generic drugs at competitive prices and maintaining quality, it caters to both domestic needs and export demands, particularly to emerging economies.

Renewable Energy: With its flat terrains and ample sunlight, Bangladesh has significant potential for solar energy. The country's solar home system program is among the world's largest, with millions of rural homes relying on solar panels for electricity.

Textiles and Jute-Based Industries: Apart from RMG, Bangladesh has a long-standing tradition of producing fine muslin fabrics and jute-based products. The diversified jute goods, including eco-friendly bags, are increasingly in demand due to global environmental consciousness.

Ceramics: Bangladesh's ceramic industry, especially tableware, tiles, and sanitary ware, has been gaining international market shares because of its high quality and competitive prices. The sector is increasingly attracting foreign investments and technology partnerships.

The extensive range of economic resources, both traditional and emerging, underscores Bangladesh's multifaceted growth potential. With the right policies and sustainable practices, these resources can collectively drive the nation towards economic prosperity.

4.5 Bangladesh's Major Economic Reforms in the Context of Political Economy

Bangladesh's trajectory since its independence in 1971 has been characterized by significant economic reforms influenced by domestic political decisions and global economic paradigms. In the context of political economy, these reforms not only reflect policy changes but also the intricate relationship between political power structures and economic considerations.

Post-Independence Nationalization (1972–1980): Following its independence, Bangladesh adopted socialist policies under Sheikh Mujibur Rahman's leadership, which included the nationalization of key industries and banks. This move aimed to consolidate state control over resources and drive equitable growth. However, inefficiencies in state-owned enterprises led to economic challenges (Ahmed 1986).

Structural Adjustment Programs (1980s): Influenced by global institutions like the IMF and World Bank, Bangladesh embarked on market-friendly reforms in the 1980s under President Ziaur Rahman and later, President Ershad. This shift involved deregulation, trade liberalization, and privatization of several state-owned entities. While these policies attracted foreign investment, critics argued they undermined social welfare and widened inequalities (Sobhan 1993).

Expansion of the RMG Sector (1980s-Onwards): Political decisions to invest and promote the ready-made garments (RMG) sector transformed Bangladesh's economic landscape. The industry, backed by favorable global trade agreements and domestic policy incentives, became the nation's primary export earner, yet it also drew criticisms regarding labor rights and environmental concerns (Kabeer and Mahmud 2004).

Financial Sector Reforms (1990s-Onwards): Emphasizing a modern and efficient financial sector, successive governments undertook reforms to strengthen regulatory frameworks, introduce private banks, and improve non-performing loan management. These reforms had mixed outcomes; while they diversified the banking sector, challenges like financial malpractices persisted (Rahman 2010).

Open Market Reforms (1990s): With the restoration of democracy in the 1990s, Bangladesh shifted towards an open market economy. Reduction of import tariffs, removal of trade barriers, and liberalized foreign direct investment policies were introduced. These measures aimed to integrate Bangladesh into the global economy, fostering an environment conducive to both local and foreign investments (Rashid 2000).

Private Sector Development Strategy (2000s): The government took action to fortify the private sector, seeing it as a potential development driver. To encourage investment from throughout the world, the Bangladesh Planning Commission recommended creating Special Economic Zones and reorganizing the Board of Investment, which is currently known as the Bangladesh Investment Development Authority (BIDA).

Energy Sector Reforms (2010s): With the country's expanding energy needs, the government began diversifying energy sources, including LNG and renewable energy. Establishing international partnerships ensured a steady energy supply for industry and households (Energy and Mineral Resources Division 2016).

Taxation and Revenue Reforms (2010s-Onwards): In order to increase revenue mobilization, tax administration and policy reforms were implemented. The Value Added Tax and Supplementary Duty Act of 2012 sought to simplify the VAT system and expand the tax base. Nonetheless, its implementation has been hindered by concerns from various economic sectors.

Export Diversification Strategy (2010s-Onwards): While RMG remains the dominant export, the government has been keen to diversify its export portfolio. Sectors like leather, pharmaceuticals, information technology, and shipbuilding have been identified as potential growth areas, and they are receiving policy support to enhance their global market presence.

The economic reforms of Bangladesh, in tandem with its political landscape, manifest the nation's adaptive strategies in navigating both domestic imperatives and global economic shifts. The trajectory underscores Bangladesh's resilience and its nuanced approach to development in the face of multifaceted challenges. Bangladesh's economic reforms, when viewed through the lens of political economy, showcase a complex interplay between political imperatives and economic considerations. The evolving nature of these reforms reflects the nation's endeavors to balance growth aspirations with socio-political realities.

4.6 Major Import and Export of Bangladesh

4.6.1 Export Products

1. **Readymade Garments (RMG)**: Bangladesh's exports still rely heavily on the RMG industry. The total value of RMG exports in FY 2021–2022 was over $31.46 billion. The United States, the United Kingdom, Germany, and other European countries are key markets for woven and knitted clothes.
2. **Jute and Jute Products**: Bangladesh is reviving its jute industry, which was once its primary export and is now the country's second largest. Raw jute exports are from 0.6 to 0.8 million bales per year, with the majority going to China, India, and Pakistan. Raw jute exports earned BDT 6.59 billion in FY 2021–22.
3. **Fish, Shrimps, and Prawns**: Bangladesh's economic success depends heavily on the aquaculture sector. According to EPB data, revenue from prawn exports was around $268.95 million in H1 FY22. Europe, the United States, and China are all major importers (Table 4.4).

4. **Leather and Leather Products**: Bangladesh's leather industry witnessed considerable growth in FY 2021–2022. Exports in leather and leather goods rose by over 30% during the first three quarters of this fiscal year, targeting markets like Italy, Spain, and Hong Kong (Table 4.5).
5. **Pharmaceutical Products**: Notably, the pharmaceutical sector grew by 27% in the first ten months of FY 2021–22. Bangladesh now exports to almost 160 countries, meeting 98% of its local pharmaceutical demands.
6. **Home Textile**: This sector experienced a 39% export increase between July 2021 and April 2022. Products range from bed linens to curtains, with the US, Europe, and Japan being key markets.

Table 4.4 Export items of Bangladesh in 2018–19 and 2019–20

Export items	2018–19 ($ in million)	2019–20 ($ in million)	% of change
Knit clothes	16,888.54	13,908.00	– 17.65
Woven clothes	17,244.73	14,041.19	- 15.58
Home textile	851.72	758.91	– 10.90
Leather, its product, shoe	1,019.78	797.60	– 21.79
Fish	500.40	456.15	- 11.52
Agricultural products	908.96	862.06	– 5.16
Pharmaceuticals	129.95	135.79	+ 4.49
Plastic, melamine, etc.	119.80	100.52	16.09
Raw jute, jute products	816.27	882.35	+ 8.09
Bi-cycle	84.28	82.84	– 1.66

Source Export Promotion Bureau 2019–20

Table 4.5 Top export destinations of Bangladesh in 2018–19 and 2019–20

Country	2018–19 ($ in million)	2019–20 ($ in million)	% of change
USA	6876.29	5832.39	– 15.18
Germany	6173.16	5099.19	– 17.40
UK	4169.13	3453.88	– 17.16
France	2217.56	1703.58	– 23.18
Spain	2554.82	2189.03	– 14.32
Italy	1643.11	1282.88	– 21.93
Canada	1339.80	1000.49	– 25.33
Belgium	946.93	723.43	– 23.60
Netherland	1278.69	1098.68	– 14.06
Japan	1365.74	1200.78	– 12.08

Source Export Promotion Bureau 2019–20

7. **Agro Products**: Vegetable exports have shown potential, with shipments doubling in the initial months of the previous fiscal year. Overall, agricultural exports brought in $1.16 billion in FY22.
8. **Plastic and Plastic Products**: Plastic exports, including toys and home decor, are on the rise. The total export value reached $128.77 million in the first ten months of the last fiscal year.
9. **Bicycle**: Targeting European markets, Bangladesh has become a significant bicycle exporter. The nation achieved an export value of $84.58 million in FY 2020–21.
10. **Terry Towel**: Focusing solely on exports, the terry towel industry generated around $18.6 million between July and December 2022.
11. **Other Items**: The burgeoning electronics sector, including mobile phones and computer components, contributes to the country's exports. Additionally, ceramic products, chemical products for textiles, furniture, and traditional crafts play a role in the export landscape.

In conclusion, Bangladesh's export portfolio showcases a diverse array of products, reflecting its economic versatility. From traditional sectors like jute to modern industries like pharmaceuticals and electronics, the nation continues to expand its global market presence.

4.6.2 Import Products

Bangladesh's economy has been growing steadily over the years, and with this growth, there's been a surge in its import requirements. As a rapidly developing country, Bangladesh relies heavily on imports to meet the demands of its population and industries. Here are the major import products of Bangladesh:

1. **Raw Cotton and Textile Machinery**: As the world's second-largest apparel exporter, Bangladesh imports massive amounts of raw cotton and textile machinery to support its garment industry. The country heavily depends on the import of raw materials for its textile sector.
2. **Petroleum and Petroleum Products**: With limited fossil fuel resources, Bangladesh imports a significant amount of its petroleum needs. Petroleum products are crucial for transportation, electricity generation, and various industrial processes (Table 4.6).
3. **Machinery and Equipment**: As industries in Bangladesh continue to expand, there's a rising demand for machinery and equipment, especially in sectors such as construction, manufacturing, and agriculture.
4. **Iron, Steel, and Cement**: The construction boom in Bangladesh, driven by urbanization and infrastructure projects, has led to increased imports of iron, steel, and cement. These materials are foundational for the country's construction and infrastructure projects.
5. **Chemicals**: Bangladesh imports a variety of chemicals used in industries ranging from textiles to pharmaceuticals. These chemicals are crucial for production processes and often lack local substitutes.
6. **Food Grains**: While Bangladesh has made significant strides in food grain production, it still occasionally relies on imports, especially during times of natural disasters or shortages. Rice and wheat are the primary food grains imported (Table 4.7).
7. **Electronic Items**: Consumer electronics, especially items like mobile phones, televisions, and computers, are primarily imported, given the limited electronic manufacturing capabilities within the country.
8. **Automobiles and Spare Parts**: With rising income levels and urbanization, there's been a surge in the demand for vehicles. While there are some local assembly lines, many automobiles, especially luxury brands, are imported, along with their spare parts.
9. **Edible Oils**: Despite efforts to increase local production, Bangladesh still relies heavily on imports to meet its edible oil requirements, including soybean oil and palm oil.
10. **Pharmaceutical Raw Materials**: Though Bangladesh's pharmaceutical industry is robust and growing, the country imports many of the raw materials required for drug production.

4.7 International Trade of Bangladesh

Bangladesh's international trade landscape has undergone significant transformations since its independence in 1971. As a Least Developed Country (LDC), the nation has historically enjoyed trade privileges that have facilitated its export-oriented growth. Post-independence, Bangladesh primarily exported raw jute and tea.

Table 4.6 Top import destinations of Bangladesh

Country	Taka (Crore)	USD ($) in million	Percentage of total
China	114,593.5	13,638.8	26.1
India	64,253.7	7647.5	14.7
Singapore	26,098.3	3105.7	6.0
Indonesia	16,305.3	1940.3	3.7
Japan	15,517.4	1846.3	3.5
USA	14,930.9	1776.6	3.4
Malaysia	12,572.4	1496.0	2.9
Brazil	11,787.7	1402.4	2.7
South Korea	11,047.2	1314.7	2.5
UAE	10,278.6	1222.6	2.3

Source Bangladesh Foreign Trade Statistics 2021–22

Table 4.7 Top import commodities of Bangladesh

Commodities	2016–17 (million USD)	2017–18 (million USD)	2018–19 (million USD)
Rice	76	1746	124
Wheat	1031	1227	1148
Milk and cream	252	288	336
Spices	259	288	315
Oil seeds	413	509	653
Edible oil	1495	1761	1718
Pulses all kinds	589	406	411
Sugar	841	951	647
Clinker	476	518	716
Crude petroleum	533	554	930

Source Bangladesh Foreign Trade Statistics 2018–19

However, with the shift in global demands and challenges in jute trade, the country gradually moved towards manufacturing, particularly in the textile and ready-made garment sectors (Bhattacharya and Rahman 2000) (Table 4.8).

The major driver behind Bangladesh's export growth is the Ready-Made Garment (RMG) sector, contributing over 80% of the nation's export revenue. Apart from RMG, Bangladesh exports jute, leather, frozen food, and agricultural products (Rahman 2019). On the other hand, Bangladesh primarily imports machinery, petroleum, raw cotton, and food grains to support its growing industries and domestic needs. The nation also imports electronic items and pharmaceutical raw materials (Table 4.9) (Fig. 4.1).

Although Bangladesh has witnessed a surge in exports, especially in RMG, the country still runs a trade deficit. This is because of its high reliance on imports

Table 4.8 Key indicators of Bangladesh's international trade (2018–2019)

Indicator	Value (in USD billion)	Reference
Total exports	39.75	EPB (2019)
Total imports	54.76	BBS (2019)
Trade balance	– 15.01	Bangladesh Bank (2019)
Major export item	RMG	Rahman (2019)
Major import item	Machinery	BBS (2019)

Source Bangladesh Foreign Trade Statistics 2021–22

Table 4.9 Trade balance of Bangladesh during last five years

Year	Trade balance (Million Tk.)	Ratio of exports and imports
2017–18	– 2,423,708	1:1.78
2018–19	– 2,241,055	1:1.64
2019–20	– 2,629,990	1:1.94
2020–21	– 3,153,162	1:1.91
2021–22	– 4,368,164	1:1.96

Source National Board of Revenue (NBR). *Note* Trade balance is negative (–) for Bangladesh

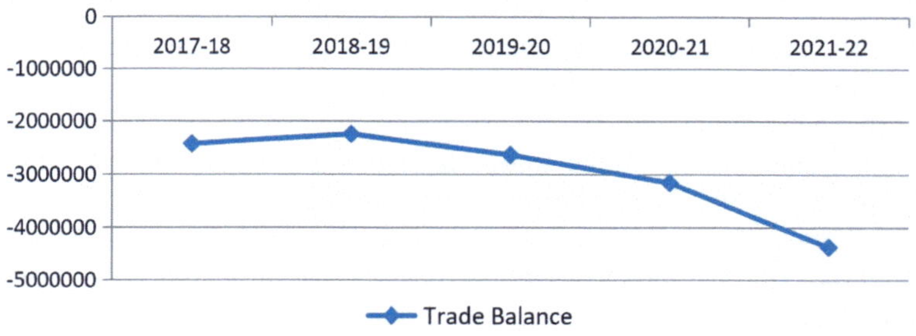

Fig. 4.1 Trade balance of Bangladesh during last five years. *Source* National Board of Revenue (NBR). *Note* Trade balance is negative (–) for Bangladesh

for essential commodities and machinery. Major Trading Partners: The European Union, especially Germany and the UK, and the USA are Bangladesh's largest export markets. For imports, China, India, and Singapore remain the top partners. Bangladesh is part of the South Asian Free Trade Area (SAFTA) and benefits from the Generalized System of Preferences (GSP) in several developed markets. The GSP advantage in the European market has been especially beneficial for the growth of its RMG sector (Rahman and Shadat 2007). Despite its successes, Bangladesh faces several challenges in international trade including dependence on a single major export item (RMG), lack of trade diversification, infrastructural bottlenecks, and compliance issues in the RMG sector (Chowdhury 2015). With the ongoing development of Special Economic Zones (SEZs) and a focus on diversifying export items, Bangladesh aims to expand its footprint in the global market. Additionally, the country's graduation from the LDC status may lead to the recalibration of trade privileges and strategies (UNCTAD 2020).

4.8 Bilateral Trade Partners of Bangladesh

Bangladesh has established substantial trade relations with numerous countries worldwide, capitalizing on its advantageous location between South and Southeast Asia. Here are the major bilateral trade partners of Bangladesh, as of recent years:

1. **China**

Nature of Trade: China is one of Bangladesh's most significant sources of imports, especially for raw materials, machinery, electronics, and consumer goods. In return, Bangladesh exports ready-made garments, leather goods, and jute products to China.

Trade Dynamics: Bangladesh's trade deficit with China is considerable due to the high volume of imports relative to exports.

Source Rahman, A. (2017). Bangladesh-China Bilateral Trade and Potential Free Trade Agreement. Bangladesh Economic Association.

2. **India**

Nature of Trade: India exports cotton, machinery, electronics, and consumer goods to Bangladesh. Bangladesh's major exports to India include textiles, jute products, seafood, and agricultural goods.

Trade Dynamics: Like China, Bangladesh also has a trade deficit with India, although initiatives have been taken to balance the trade.

Source: Taneja, N., & Bimal, S. (2016). India-Bangladesh Bilateral Trade and Potential Free Trade Agreement. Indian Council for Research on International Economic Relations.

3. **United States**

Nature of Trade: The USA primarily imports ready-made garments, footwear, headgear, textiles, and agricultural products from Bangladesh.

Trade Dynamics: The United States stands as one of the largest single-country destinations for Bangladeshi exports.

4. **European Union (EU)**

Nature of Trade: The EU, especially countries like Germany, the UK, France, and Italy, heavily imports ready-made garments, textiles, and seafood products from Bangladesh.

Trade Dynamics: The EU grants Bangladesh duty-free and quota-free access for its products under the Everything but Arms (EBA) initiative, boosting Bangladeshi exports to the region.

5. **Japan**

Nature of Trade: Japan imports textiles, leather, and seafood from Bangladesh. In return, Bangladesh imports machinery, vehicles, and electronic equipment from Japan.

Trade Dynamics: Both countries have been strengthening their trade relations, with Bangladesh aiming to expand its exports to Japan.

6. **ASEAN Countries**

Nature of Trade: With countries like Malaysia, Thailand, and Singapore, Bangladesh trades in textiles, agricultural products, and seafood. In return, Bangladesh imports machinery, electronics, and petroleum products.

Trade Dynamics: To promote and expand trade interactions, bilateral trade agreements and economic cooperation are in place.

The powerful ready-made garment industry in Bangladesh is a major factor in the country's trade dynamics. Although the country's commerce with the EU is generally advantageous, its trading with China and India is problematic due to large trade deficits. Maintaining a healthy trade balance and expanding exports are critical for long-term economic expansion.

4.9 Bangladesh in Multilateral Economic Forum

Bangladesh's presence at these international economic meetings is evidence of the country's rising stature on the world stage. When it comes to international economic talks and decision-making, the country has evolved from a passive recipient to an active participant throughout the years. In this paper, we'll look at Bangladesh's participation in and contributions to several important multilateral economic forums, such as the United Nations, the World Trade Organization, the South Asian Association for Regional Cooperation, and the Bay of Bengal Initiative for Multi-Sectoral Technical and Economic Cooperation (BIMSTEC).

1. **United Nations (UN)**

Following its independence in 1971, Bangladesh has been an active member of the United Nations since 1974. The nation has consistently advocated for the interests of developing nations on UN economic platforms such as the United Nations Conference on Trade and Development and the Economic and Social Council. Bangladesh has aligned its national development objectives with the Sustainable Development objectives (SDGs) of the United Nations (Rahman 2017).

2. **World Trade Organization (WTO)**

Bangladesh has played a significant role in advocating for the rights of Least Developed Countries (LDCs) within the WTO since its accession in 1995. The country has played a significant role in negotiations on a variety of issues, including trade-related aspects of intellectual property rights (TRIPS) and the Generalized System of Preferences (GSP), pressing for favorable terms for developing economies.

3. **South Asian Association for Regional Cooperation (SAARC)**

Since its inception in 1985, the South Asian Association for Regional Cooperation (SAARC) has counted Bangladesh among its eight member states. Food security, rural development, and energy cooperation are just a few of the regional issues that Bangladesh has spearheaded regional initiatives on as host of multiple SAARC summits.

4. **Bay of Bengal Initiative for Multi-sectoral Technical and Economic Cooperation (BIMSTEC)**

BIMSTEC acts as a bridge between South and Southeast Asia and signifies a strengthening of relations between these nations. As a founding member of BIMSTEC, Bangladesh has prioritized regional connectivity, counterterrorism, climate change, and technological innovation.

5. **World Bank and International Monetary Fund (IMF)**

Bangladesh's relationships with both the World Bank and the IMF are enduring. Over the years, the World Bank has provided Bangladesh with substantial financial and technical assistance for a variety of development initiatives, from infrastructure to health. On the other hand, the IMF has engaged in economic consultations, financing arrangements, and capacity development activities with Bangladesh in an effort to promote macroeconomic stability (IMF 2018).

6. **Commonwealth of Nations**

Bangladesh has collaborated with other Commonwealth members on trade, development, and governance issues since it joined in 1972. Bangladesh has been able to participate in initiatives focused on sustainable development, democracy, and peace due to the Commonwealth's shared values and principles (Commonwealth Secretariat 2017).

7. **The Asia–Pacific Trade Agreement (APTA)**

Bangladesh is a prominent member of APTA, the Asian preferential trade agreement. Through this agreement, Bangladesh has pursued trade liberalization measures, increasing market access for its exports, especially in the textile and apparel sectors (APTA 2016).

8. **Organization of Islamic Cooperation (OIC)**

Bangladesh is an active OIC member due to its Muslim majority. Bangladesh has promoted economic cooperation among member nations, especially the Islamic financial market and halal industry (OIC 2018).

Bangladesh's participation in multilateral economic forums shows its dedication to developing nations and regional and worldwide economic cooperation. The government will play a larger role in determining these multilateral institutions' economic policies and objectives.

9. **Developing-8 (D-8) Organization for Economic Cooperation**

The Developing-8 (D-8) is an organization of eight developing countries with considerable Muslim populations. Established in 1997, its members include Bangladesh, Egypt, Indonesia, Iran, Malaysia, Nigeria, Pakistan, and Turkey. The main goal of D-8 is to boost member countries' economic cooperation and global standing. Bangladesh is important in D-8. Bangladesh, a founding member, has promoted trade, industry, agriculture, and energy resources among member states.

4.9.1 Economic Dependency and External Political Influence in Bangladesh: The Dynamics of USA, China, and India

Bangladesh, being classified as a developing country, has exhibited a growing inclination towards incorporating itself inside the global economic framework. The nation's strategic geopolitical position, situated between the two prominent Asian powers of India and China, along with its expanding economic ties with influential global actors such as the United States, has rendered it a significant nexus of geopolitical and economic concerns. This study investigates the economic interdependencies of Bangladesh and their implications for external political influences, with a specific focus on the United States, China, and India.

USA: Trade and Development Aid

Throughout history, the United States has consistently emerged as one of the primary destinations for Bangladesh's exports, with a special emphasis on textiles and apparel. In addition, through USAID and other agencies, the United States has provided Bangladesh with substantial development assistance for health, education, and infrastructure initiatives. While this economic relationship has boosted Bangladesh's economy, it also introduces political influences from the United States, particularly in the areas of labor rights, governance, and strategic alliances.

China: Infrastructure Investment and Trade

Infrastructure in Bangladesh shows how China's economic link with the country is growing. China has spent a lot of money building ports, highways, and power plants in Bangladesh as part of the Belt and Road Initiative (BRI). These investments stimulate the economy greatly, but they also have political ramifications. Some fear that China is exerting undue influence on Bangladeshi politics and foreign policy due to the country's growing indebtedness to China and the potentially game-changing nature of certain infrastructure projects.

India: Historical Ties and Security Concerns

India played a crucial role in Bangladesh's independence and has since maintained strong economic ties with the country. The shared border and cultural history mean that trade, water resources, and migration are perennial issues. India's concern over security, especially related to insurgent groups, has often influenced its economic and political interactions with Bangladesh. Thus, while Bangladesh benefits from its economic relationship with India, it also navigates a complex political landscape influenced by historical and security concerns.

Bangladesh's economic interactions with the USA, China, and India are crucial for its development trajectory. However, these economic dependencies also bring with them external political influences. As Bangladesh continues to grow and assert itself on the global stage, navigating these relationships while safeguarding its national interests will be of paramount importance.

4.9.2 Role of Mega Development Project and EPZ in the Economy of Bangladesh

Bangladesh, over the past few decades, has witnessed significant economic growth, much of which can be attributed to strategic development initiatives, notably mega development projects and Export Processing Zones (EPZs). These initiatives have not only spurred economic activities but also positioned Bangladesh as an attractive destination for foreign direct investment (FDI) and trade.

4.9.3 Mega Development Projects

The Padma Bridge, the Rooppur Nuclear Power Plant, and the Dhaka Metro Rail, among others, are examples of mega development initiatives in Bangladesh. These initiatives have wide-ranging effects on the economy:

Economic Stimulus: Such projects generate employment, both directly within the projects and indirectly in ancillary industries and services. For instance, the construction of the Padma Bridge is estimated to add 1% to the nation's GDP.

Connectivity and Integration: Projects like the Dhaka Metro Rail alleviate transportation bottlenecks, facilitating smoother movement of goods and people, thereby potentially reducing transaction costs and boosting intra-country trade (Rahman 2019).

Attracting FDI: Major infrastructural developments signal a country's commitment to modernization, making it a more attractive proposition for foreign investors.

Enhancing Regional Connectivity: Mega development projects, particularly in transportation and energy, have the potential to enhance regional connectivity. The Padma Bridge, for example, is anticipated to strengthen connections between the eastern and western parts of South Asia, paving the way for greater regional trade and cooperation (Islam 2021).

Environmental and Sustainability Concerns: Notably, while megaprojects stimulate economic development, they have also raised environmental concerns. To ensure that economic growth does not occur at the expense of the environment, proper environmental impact assessments and sustainable practices are crucial (Ahmed 2020).

4.9.4 Export Processing Zones (EPZs)

EPZs are designated areas in Bangladesh aimed at promoting exports by offering a conducive environment for businesses through regulatory, tax, and infrastructural incentives. The impact of EPZs on the Bangladeshi economy has been profound:

Boosting Exports: EPZs, primarily driven by the garment industry, have contributed significantly to Bangladesh's export earnings. In 2019 alone, EPZs contributed to approximately 20% of the country's total exports.

Employment Generation: EPZs have created millions of jobs, especially for women, leading to socio-economic upliftment and increased disposable income levels in families.

Transfer of Technology and Skills: Foreign companies in EPZs often bring advanced technologies and practices, leading to skill enhancement of the local workforce.

Diversification of the Economy: EPZs have allowed Bangladesh to diversify its economic base. While the garment industry remains dominant, EPZs have attracted investments in electronics, pharmaceuticals, and information technology, reducing the country's vulnerability to global market fluctuations in any single sector (Khan 2019).

Role in Urbanization: Both mega projects and EPZs have accelerated the pace of urbanization in Bangladesh. They attract a workforce from rural areas, leading to rapid urban growth. This has implications for urban planning, housing, and municipal services (Rahman and Hossain 2018).

Governance and Transparency: The scale and significance of mega projects and EPZs underline the need for good governance and transparency. Effective monitoring and accountability mechanisms ensure that these initiatives deliver their promised economic benefits and minimize potential malpractices (Hasan 2019).

Strengthening Bilateral Relations: Economic engagements through EPZs often lead to stronger bilateral relations with countries that invest heavily in these zones. This can further open up avenues for diplomatic collaborations and partnerships in other sectors, such as defense, education, and cultural exchanges (Biswas 2020).

Mega development projects and EPZs are pivotal in reshaping the economic landscape of Bangladesh. While the former focuses on enhancing infrastructure and connectivity, promoting economic activities and attracting FDI, the latter specifically aims to bolster the nation's export capabilities. Together, they represent a dual strategy of internal development and external economic engagement, driving Bangladesh's journey towards becoming a middle-income country.

4.9.5 Economic Diplomacy: Bangladesh Perspective

The three pillars of Bangladesh's economic diplomacy are Foreign Direct Investment (FDI), the export of diverse commodities, and overseas Bangladeshi laborers. Unlike traditional diplomacy, this economic strategy isn't solely executed by diplomats abroad but is a collaborative effort involving ministers, bureaucrats, business stakeholders, and various governmental agencies at home. Additionally, non-state actors like NGOs, businesspersons, and investors play an essential role. However, the efficacy of Bangladesh's economic diplomacy faces obstacles, including infrastructural weaknesses, bureaucratic hurdles, and issues of transparency and coordination. Regarding FDI, feedback from potential investors underscores the intricate, drawn-out process in Dhaka, which often dissuades investments. An optimal approach should ensure preliminary information and guidelines are available both online and at overseas missions. A central committee could expedite decision-making on investment proposals. Automatic approvals, with certain sectoral exclusions, could also streamline smaller projects.

Diversifying exports is paramount. While Bangladesh has a handful of key exportable items, expanding this basket is essential to meet global demand. Exploring new verticals, especially technology, pharmaceuticals, and agriculture, could bolster exports. For overseas workers, skills enhancement is non-negotiable. A skilled workforce ensures better acceptability and compensation in host nations. The pandemic poses fresh challenges for migrants, necessitating urgent attention. Furthermore, leveraging strengths in areas like disaster management could pave new avenues in economic diplomacy, akin to initiatives by Brazil's BCA or India's DPA. Ultimately, while robust policies are vital, successful economic diplomacy hinges on projecting a positive, compelling image of Bangladesh on the global stage.

4.9.6 Economic Diplomacy of Bangladesh: A Strategic Framework for Development

As an integral component of Bangladesh's foreign policy, economic diplomacy has emerged as a potent instrument for promoting economic development, trade, and investment on the international stage. In a rapidly changing international landscape, Bangladesh has strategically employed economic diplomacy to secure its national interests, strengthen international partnerships, and capitalize on economic opportunities. This paper explores the framework and key elements of Bangladesh's economic diplomacy, emphasizing its significance in the country's development trajectory.

4.9.7 Economic Diplomacy: An Evolving Paradigm

Economic diplomacy encompasses a range of activities aimed at advancing economic interests through diplomatic means. For Bangladesh, economic diplomacy has evolved into a comprehensive strategy that leverages its position as a growing economy with a strategic location in South Asia. The country has recognized that economic development is not confined within its borders but also hinges on international engagement.

4.9.8 Key Elements of Bangladesh's Economic Diplomacy

Trade Facilitation and Promotion

Bangladesh's diplomatic missions worldwide actively promote trade by facilitating exports, attracting foreign investment, and negotiating trade agreements. Notably, the Ministry of Commerce's initiatives focus on diversifying export products and markets, with a particular emphasis on the USA, China, and India.

Investment Promotion

Bangladesh Economic Zones Authority (BEZA) and the Board of Investment (BOI) play pivotal roles in attracting foreign direct investment (FDI) by offering incentives, streamlining investment procedures, and providing one-stop services for investors.

Multilateral Engagement

Bangladesh participates actively in multilateral economic forums such as the United Nations, the World Trade Organization, and regional organizations such as SAARC and BIMSTEC. These forums offer opportunities to advocate for the interests of developing nations, particularly during trade negotiations.

Regional Connectivity

Improving regional connectivity through infrastructure development and regional trade agreements, like the BBIN initiative, has been a key focus. Such initiatives enhance Bangladesh's role as a transit hub and foster economic integration with neighboring countries.

Market Access and Preferential Trade Agreements

Bangladesh is actively involved in preferential trade agreement (PTA) and regional trade bloc negotiations. Bangladesh has benefited from PTAs with numerous countries and regions, including the European Union's Everything But Arms (EBA) initiative, which grants duty-free and quota-free access to the EU market for Bangladeshi products. In addition to providing market access, these agreements foster diplomatic relations and regulatory collaboration.

Bilateral Economic Relations

Economic relations between two countries are very important for Bangladesh's economic growth. The country has business ties with many other countries, and its diplomatic missions around the world actively support trade and investment between the two countries. Bangladesh's economic links with Japan, for example, have led to large amounts of Japanese investment in textiles and manufacturing.

Economic Diplomacy for Sustainable Development

Bangladesh's economic diplomacy also encompasses efforts to attract green and sustainable investments. The country recognizes the importance of environmentally responsible economic activities and seeks investments and partnerships aligned with sustainability goals. This includes projects related to renewable energy, green infrastructure, and sustainable agriculture.

Economic Diplomacy for Human Capital Development

Bangladesh's economic diplomacy places a premium on investing in human capital. It aims to work with other nations and international organizations to increase cooperation and the sharing of knowledge in fields such as technology, education, and healthcare. Collaborations with nations like South Korea in training and education programs are evidence of the efficacy of this strategy.

Diaspora Engagement

Bangladesh has an extensive diaspora abroad, and it actively engages with the Bangladeshi expatriate community to advance economic diplomacy. This includes pursuing investment opportunities, encouraging remittances, and leveraging the expertise and networks of the diaspora to strengthen economic connections.

Investment Promotion Agencies

To welcome and safeguard international investment, Bangladesh has set up government bodies such as the Investment Promotion and Protection Agency (BIPA).

These organizations are essential in fostering a welcoming environment for investors, disseminating useful data, and settling conflicts that arise from investments.

Economic Diplomacy in Crisis Management

Bangladesh's economic diplomacy includes soliciting foreign support during crises like natural catastrophes and the COVID-19 pandemic. For humanitarian and economic issues, the government has aggressively sought aid and collaboration.

Bangladesh has benefited from economic diplomacy by increasing exports, FDI, and resilience during global crises. It also struggles with trade imbalances, policy coherence, and economic-related environmental issues. Bangladesh's economic diplomacy now uses its economic resources and geopolitical location to go beyond traditional diplomacy. Economic diplomacy is essential for sustainable economic growth, poverty reduction, and human well-being as the country moves towards middle-income status.

References

Alamgir M (1972) Famine in South Asia. Oelgeschlager, Gunn & Hain

Ahmed M (1986) Political economy of Bangladesh. University Press Limited, Dhaka

Ahmed F (2020) Mega development projects and environmental implications: a study on Bangladesh. Environ Develop J 12(1):20–34

Ahmed F, Hossain M (2018) Infrastructure development and economic growth in Bangladesh. J Econ Struct

Asia-Pacific Trade Agreement (2016) Benefits of the Asia-Pacific trade agreement for member states. Retrieved from APTA website

Berik G, Rodgers YV (2008) Bangladeshi women's labor force participation and conditions of work in readymade garment factories. Dev Chang 39(6):1239–1266

Bhattacharya D, Rahman M (2000) Female employment under export-propelled industrialization: prospects for internalizing global opportunities in Bangladesh's Apparel Sector. UNRISD

Biswas R (2020) EPZs and diplomacy: strengthening bilateral relations through economic partnerships. Diplomatic Rev 18(1):38–52

Chowdhury A (2017) The Belt and road initiative: implications for Bangladesh. Institute of Strategic Studies Islamabad

Chowdhury MMH (2015) RMG: prospects and challenges and role of different stakeholders. The Financial Express

Commonwealth Secretariat (2017) Bangladesh and the commonwealth. Retrieved from Commonwealth website

Energy and Mineral Resources Division (2016) Bangladesh Energy Regulatory Commission Annual Report. Government of Bangladesh, Dhaka

Hasan T (2019) Governance in mega projects: the Bangladesh experience. Governance Policy Rev 10(4):45–60

International Monetary Fund (2018) Bangladesh: IMF Country Report. Retrieved from IMF website

Islam R (2021) Padma bridge and the promise of regional connectivity. Dhaka South Asia Regional Stud J 7(2):46–59

Kabeer N, Mahmud S (2004) Globalization, gender and poverty: Bangladeshi women workers in export and local markets. J Int Dev 16(1):93–109

Khan A (2019) Diversifying the economic base: role of EPZs in Bangladesh. Econ Perspect 24(3):78–93

Ministry of Commerce, Bangladesh (2020) Export Policy 2020–2023
Organization of Islamic Cooperation. (2018) Bangladesh and the OIC: economic opportunities. Retrieved from OIC website
Osmani SR (2004) The Macroeconomics of poverty reduction: the case of Bangladesh. UNDP
Paul BP, Rashid S (2018) Natural hazards in Bangladesh: mainstreaming adaptation and resilience for sustainable development. Springer, Berlin
Rahman HZ (2010) The political economy of the People's Republic of Bangladesh. Centre for Policy Dialogue, Dhaka
Rahman A (2017) Bangladesh-China bilateral trade and potential free trade agreement. Bangladesh Economic Association
Rahman M (2019) Evolution of the ready-made garment industry in Bangladesh: changes in the workforce composition and implications for worker rights and gender equality. Asian J Women's Stud
Rahman M, Hossain N (2018) Urbanization in Bangladesh: challenges and opportunities. Urban Stud J 15(2):112–127
Rahman M, Shadat WB (2007) Navigating the contours of Bangladesh Trade Policy: analyzing performance, identifying problems and suggesting some policy actions. Centre for Policy Dialogue.
Rashid MA (2000) Trade policies, reforms and development: an analysis of Bangladesh's Decade of Liberalisation. Bangladesh Institute of Development Studies, Dhaka
Siddiqui T, Abrar CR (2003) Migrant worker remittances and micro-finance in Bangladesh. International Labour Organization
Sobhan R (1993) Political economy of the state and market in Bangladesh. University Press Limited, Dhaka
Taneja N, Bimal S (2016) India-Bangladesh bilateral trade and potential free trade agreement. Indian Council for Research on International Economic Relations
Transparency International (2019) Corruption perceptions index: Bangladesh
UNCTAD. (2020). The Least Developed Countries Report 2020
Zaman K, Akhter S (2018) Challenges to the sustainable growth of Bangladesh RMG. J World Econ Res

Chapter 5
Trade Data and Analysis: Perspective Between Bangladesh and LAC

Abstract The chapter explores the evolving economic ties between Bangladesh and Latin American Countries (LAC), highlighting mutual opportunities, existing challenges, and potential areas for enhanced cooperation. Despite geographical distance and historically limited engagement, both regions are increasingly recognizing the strategic importance of diversifying their trade relationships. Bangladesh's export economy, dominated by the ready-made garments (RMG) sector, continues to drive its global trade. Recent data reveal exports to LAC reaching approximately TK 44.56 billion in 2021–22, led by products such as textiles, jute goods, and footwear. In contrast, Bangladesh's imports from LAC surged to around TK 430.98 billion during the same period, reflecting heavy reliance on agricultural commodities, including soybeans from Brazil and wheat from Argentina. This trade imbalance underscores a significant dependency on LAC's resource-driven economies for raw materials and food security. Brazil remains Bangladesh's largest trading partner in the region, accounting for nearly 50% of imports, primarily agricultural and mineral products. Chile emerges as a leading export destination, with steady demand for Bangladesh's RMG products. Smaller economies such as Paraguay, Uruguay, and Peru also contribute to the trade ecosystem, albeit at varying scales. However, countries like Bolivia and French Guiana exhibit minimal trade volumes, signaling untapped potential. Analyzing trade trends over the past three decades reveals rapid growth in bilateral trade but also consistent barriers such as logistical inefficiencies, lack of direct shipping routes, and limited awareness of market-specific demands. The chapter identifies emerging sectors like pharmaceuticals, IT services, and eco-tourism as areas where bilateral engagement can be expanded, benefiting from Bangladesh's skilled labor force and LAC's diversified natural resources. Data from 2019 to 2022 illustrate marked fluctuations influenced by global disruptions, including the COVID-19 pandemic. Exports showed resilience, rebounding by 50% from 2020 to 2022. Imports, however, witnessed a sharper increase, driven by escalating demand for LAC's agricultural and mineral exports. This trend raises questions about Bangladesh's efforts towards self-sufficiency and trade diversification. The chapter further delves into the socio-economic and environmental dimensions of this relationship, emphasizing the role of sustainability and green technologies in shaping future trade patterns. Recommendations include fostering stronger regional

M. T. Islam and M. B. Hossen, *Economic Diplomacy*, The Latin American Studies Book Series, https://doi.org/10.1007/978-3-031-81362-7_5

alliances, leveraging trade agreements, and investing in infrastructural upgrades to reduce trade costs. Bangladesh and LAC nations must align on shared goals like value addition, innovation, and climate resilience to build a more balanced and mutually beneficial partnership. Economic diplomacy between these regions has begun to address structural barriers, such as logistical challenges, limited market awareness, and lack of direct trade routes. Initiatives like bilateral trade delegations, participation in global trade forums, and strategic engagement through multilateral platforms such as WTO are highlighted as critical steps towards reducing trade frictions. Furthermore, the chapter examines the role of cultural and geopolitical considerations, noting how soft power and diplomatic missions can enhance mutual understanding and trust. The analysis also highlights the strategic implications of this partnership. For Bangladesh, engaging with LAC offers access to critical raw materials, alternative markets, and geopolitical diversification. For LAC nations, strengthening ties with Bangladesh provides a gateway to Asia's broader economic landscape. By emphasizing economic diplomacy, both regions can reshape their trade relations into strategic alliances, addressing global challenges while enhancing regional and global influence. The chapter presents a nuanced perspective on the growing role of economic diplomacy in reshaping Bangladesh–LAC relations. By transcending transactional trade dynamics, these nations have the opportunity to build a partnership characterized by innovation, sustainability, and strategic collaboration, positioning themselves as key players in a rapidly evolving global economy.

Keywords Bangladesh · LAC · International trade · Imports · Exports · Economic diplomacy

5.1 Introduction

The examination and interpretation of trade data and analysis play a pivotal role in understanding the economic dynamics between countries, granting valuable perspectives on their respective competitive strengths, market entry opportunities, and overall economic approaches. The trade relationship between Bangladesh and the countries of Latin America is a topic that has received relatively less attention; however, it is becoming increasingly significant. Bangladesh has gained international recognition for its notable economic expansion and flourishing textile sector. Conversely, the LAC region is commonly linked to economies abundant in resources, with a primary emphasis on agricultural and mineral exports. Nevertheless, both regions are actively endeavoring to broaden their economic portfolios and participate in activities that provide value, rendering their trade partnership notably significant.

Historically, Bangladesh has exhibited a tendency to priorities commercial relations with Western nations and neighboring economies in South Asia. However, in recent years, there has been a notable change in the country's approach as it endeavors to explore alternative markets in order to maintain its growth driven by exports. Similarly, LAC countries are keen to diversify their trading partners beyond their historical focus on the United States and European nations, to counter volatility in these markets and to tap into the high-growth Asian markets (ECLAC 2020a). Preliminary analyses suggest that there is considerable scope for mutual benefit. Bangladesh can offer LAC countries cost-effective manufacturing and a young, skilled labor force, while LAC can offer Bangladesh an array of raw materials, particularly in agriculture and mining, that are scarce domestically (Ahmed and Gutierrez 2019). Yet, the trade volume between Bangladesh and LAC remains suboptimal, pointing to the existence of barriers that range from logistical issues to a lack of awareness of market opportunities. Understanding the trade dynamics between Bangladesh and LAC is essential for policymakers, businesses, and academics alike. Comprehensive trade data and analyses can facilitate the identification of untapped opportunities, and help both regions to better position themselves in the global economic landscape. This also aligns with broader economic development goals, such as the United Nations' Sustainable Development Goals, particularly in promoting international trade and cooperation (UN 2015). The methodological approach for analyzing trade between Bangladesh and LAC is multi-faceted and interdisciplinary, drawing on trade theories such as comparative advantage (Ricardo 1817) and the gravity model of trade (Tinbergen 1962), as well as incorporating recent advancements in network analysis (Newman 2010). This is critical, as the dynamism of these trading relationships extends beyond mere economic metrics and enters the realm of social and political considerations (Hirschman 1958). For example, how do geopolitical strategies and cultural influences shape this trade relationship? And what role do international organizations like the World Trade Organization (WTO) and regional trade agreements play in facilitating or hindering such interactions (Baldwin and Low 2009)? Furthermore, a critical but often overlooked aspect is the role of Small and Medium-sized Enterprises (SMEs) in shaping this trade relationship (Ayyagari et al. 2011). Bangladesh has a burgeoning SME sector that could find willing partners and niche markets in the LAC region, particularly in the technology and service sectors (Ali and Hatta 2019). Similarly, LAC's SMEs in sectors like eco-tourism and artisanal crafts could find a new consumer base in Bangladesh (Gomez and Santamaria 2020). Moreover, in an era characterized by increasing concerns about sustainability, both Bangladesh and LAC countries are grappling with the impacts of climate change on their trade logistics and agricultural productivity (Stern 2007). It is therefore imperative to analyze trade data through the lens of environmental sustainability and understand how green technologies can be part of the trading portfolio between these regions (Brown and Stern 2018). In sum, the multifaceted nature of trade between Bangladesh and the LAC region demands a comprehensive and interdisciplinary approach for analysis. The integration of various theories, sectoral analyses, and sustainability considerations promises to offer nuanced and actionable insights that can guide both policy and practice in shaping a more fruitful trading relationship. As Bangladesh and LAC countries aim to climb the value-added chain and diversify their trade portfolios, an academic scrutiny of their trade data and patterns is timely and can provide actionable insights. This study, therefore, endeavors to fill this gap in the literature by offering a nuanced perspective on the trade relationship between Bangladesh and the LAC region.

5.2 Statistics of International Trade for Bangladesh

The landscape of international trade for Bangladesh has evolved remarkably over the years, presenting a compelling study in economic resilience and diversification. As a cornerstone of the national economy, international trade has catapulted Bangladesh into being a significant global player, particularly in the textile and garment sectors.

According to the most recent data, Bangladesh's exports stood at approximately $40.53 billion for the fiscal year 2019–2020, primarily driven by its robust textile industry, which accounts for more than 80% of total exports. On the flip side, imports were valued at around $55.44 billion for the same period, with a diversified portfolio that includes machinery, chemicals, and foodstuffs. Despite a consistent trade deficit—around $14.91 billion in the fiscal year 2019–2020—the nation is on a steadfast path to diversify its export base and explore new markets. The dataset (Table 5.1) presents an illuminating view of Bangladesh's international trade dynamics over the past three decades, from 1992–93 to 2021–22. A cursory glance reveals some striking patterns:

Rapid Growth in Trade Volumes: Both export and import numbers have seen a remarkable uptick over the years. While the export figures were a modest 88,215 (in unspecified units) in 1992–93, they skyrocketed to 4,571,354 by 2021–22. A similar trend is observed in imports, with numbers ballooning from 138,198 in 1992–93 to a staggering 8,939,518 in 2021–22.

Persistently Negative Trade Balance: Despite the burgeoning trade volumes, Bangladesh has consistently posted a negative trade balance, indicating that imports have outstripped exports every year. Interestingly, the trade deficit has also been widening over the years, peaking at – 4,368,164 in 2021–22, far higher than the – 49,983 recorded in 1992–93.

Boom and Bust Cycles: While growth has generally been consistent, certain years like 2008–09 and 2019–20 show a noticeable dip in exports, which might be attributable to global economic downturns, like the financial crisis of 2008 and the COVID-19 pandemic in 2019–2020.

Volatile Dynamics: The numbers indicate trade balance volatility in the early 2000s and after 2010. Economic policy changes, global demand changes, and geopolitical events might affect these swings.

Trend analysis, regression modelling, and economic forecasting can reveal the country's economic health and trade policies from this data. It guides policymakers and academics through Bangladesh's complex international trade dynamics using a statistical roadmap (Table 5.2).

Table 5.1 Summary of exports, imports, and balance of trade during the last 30 years (million Tk.)

Year	Export	Import	Balance
1992–93	88,215	138,198	– 49,983
1993–94	98,739	137,540	– 38,801
1994–95	136,970	218,564	– 81,594
1995–96	144,521	254,646	– 110,125
1996–97	171,554	290,187	– 118,633
1997–98	229,408	318,916	– 89,508
1998–99	245,620	341,017	– 95,397
1999–00	247,415	372,022	– 124,606
2000–01	306,477	436,949	– 130,472
2001–02	302,119	518,443	– 216,324
2002–03	337,885	506,200	– 168,315
2003–04	437,098	630,363	– 193,265
2004–05	532,831	769,954	– 237,123
2005–06	691,950	962,345	– 270,395
2006–07	850,309	1,118,664	– 268,355
2007–08	985,931	1,496,722	– 510,791
2008–09	1,074,992	1,580,898	– 505,906
2009–10	1,134,589	1,794,324	– 659,735
2010–11	1,629,733	2,659,405	– 1,029,672
2011–12	1,762,288	2,942,810	– 1,180,522
2012–13	2,268,607	3,144,291	– 875,684
2013–14	2,122,602	2,822,310	– 699,708
2014–15	2,408,850	3,670,702	– 1,261,852
2015–16	2,634,668	3,869,349	– 1,234,681
2016–17	3,003,837	4,712,495	– 1,708,658
2017–18	3,087,936	5,511,644	– 2,423,708
2018–19	3,481,620	5,722,675	– 2,241,055
2019–20	2,811,668	5,441,658	– 2,629,990
2020–21	3,479,890	6,633,052	– 3,153,162
2021–22	4,571,354	8,939,518	– 4,368,164

Source Bangladesh Foreign Trade Statistics 2020–21

5.3 Export and Import of Commodities by Section

Table 5.3 provides a comprehensive analysis of Bangladesh's export performance during six consecutive fiscal years, encompassing measurements in both the local currency (Tk.) and US dollars.

Here's an analysis of the data: Yearly Export Growth in Tk:

Table 5.2 Sectional export and import of commodities in 2019–20, 2020–21, and 2021–22 (million Tk.)

Section	Commodity	2019–20		2020–21		2021–22	
		Export	Import	Export	Import	Export	Import
1.	Live animal; animal products	36,482	51,636	42,942	51,842	47,949	69,479
2.	Vegetable products	22,106	442,319	21,800	616,037	22,829	666,704
3.	Animal/veg. fat/oil, edible fat, wax	10,990	240,035	16,628	305,645	19,858	461,881
4.	Prepare food, beverage, spirit, tobacco	34,047	184,550	49,667	232,775	56,400	307,289
5.	Mineral products	14,817	887,265	8449	1,319,083	7791	1,388,112
6.	Product chemical or allied industry	15,630	422,097	24,612	516,426	31,433	988,518
7.	Plastic, rubber and articles thereof	10,438	255,924	13,473	306,208	18,694	399,617
8.	Raw hide and skin, leather, fur skin	25,898	16,750	36,983	16,604	42,285	23,366
9.	Wood and articles, wood, cork, basket	1852	11,913	403	10,744	470	14,305
10.	Pulp wood, celluloid, paper/board	6092	96,935	9326	96,372	12,968	129,133
11.	Textile and textile articles	2,428,761	11,746,552	3,017,230	1,296,116	3,989,856	1,986,512
12.	Footwear, headgear, umbrellas, feather	79,231	19,557	104,309	21,096	144,576	37,294
13.	Article stone, plaster, cement, glass	2553	28,044	3477	34,461	5296	37,830
14.	Natural/cultured pearl, precious stone	22	715	61	1322	66	1382

(continued)

Table 5.2 (continued)

Section	Commodity	2019–20		2020–21		2021–22	
		Export	Import	Export	Import	Export	Import
15.	Base metals and articles base metal	9111	466,969	19,930	482,801	24,460	744,193
16.	Machine/ mechanic apparel, electric equip	49,779	755,752	44,214	920,389	46,854	1,217,906
17.	Vehicle, aircraft, vessel/trans. equip	39,676	276,094	34,898	294,750	64,560	329,415
18.	Optical, photographic, surgical instrument	9739	54,169	11,150	52,543	11,393	64,225
19.	Arms ammunition and parts thereof	1	13,404	6764	11,896	9556	8288
20.	Miscellaneous manufactured articles	14,422	42,782	13,544	45,181	14,020	63,988
21.	Work of art, collectors pieces	21	96	30	761	40	81

Source Bangladesh Foreign Trade Statistics 2020–21

Table 5.3 Exports of Bangladesh during last six years

Year	Exports (million Tk.)	Exports (million US dollar)	% Change (in Tk.)
2016–17	3,003,837	37,966	14.01
2017–18	3,087,936	37,612	2.80
2018–19	3,481,620	41,433	12.75
2019–20	2,811,668	33,164	– 19.24
2020–21	3,479,890	41,032	23.77
2021–22	4,571,354	52,970	31.36

Source Bangladesh Foreign Trade Statistics 2020–21

2016–17: A growth of 14.01% depicts a solid year for exports.

2017–18: There was a slowdown in growth to just 2.80%, indicating a possible stagnation or challenges faced in expanding the export market.

2018–19: Exports rebounded with a commendable 12.75% increase.

2019–20: A significant decline was observed this year, with exports dropping by 19.24%. This negative growth might be attributed to external factors like the global economic slowdown, trade challenges, or internal factors affecting production and trade.

2020–21: The country managed to recover remarkably, with exports surging by 23.77%.

2021–22: This year saw the highest growth in the given period, with exports booming by 31.36%, signifying a robust recovery and expansion phase.

Exports in US Dollars: While the Tk. value of exports gives a general sense of growth, the US dollar figures are crucial for understanding the performance in an international context. Notably, the highest export value in US dollars was achieved in 2021–22 at $52,970 million, suggesting a strong global demand for Bangladeshi goods and/or a favorable exchange rate scenario.

External Factors: The decline in 2019–20 might correlate with global events such as the onset of the COVID-19 pandemic, which disrupted global trade and supply chains. The subsequent years' recovery indicates Bangladesh's resilience and adaptability in navigating global challenges (Tables 5.4 and 5.5).

Over the last six years, Bangladesh's export sector has showcased resilience, adaptability, and an upward trajectory, especially in the face of global challenges. The increasing export figures, both in terms of Tk. and US dollars, are indicative of the country's growing significance in the global trade landscape.

Table 5.4 Imports of Bangladesh from 2017–18 to 2021–22

Year	Imports (million Tk.)	Imports (million US dollars)	% Change (In Tk.)
2017–18	5,511,644	67,133	16.96
2018–19	5,722,675	68,103	3.83
2019–20	5,441,658	64,186	− 4.91
2020–21	6,633,052	78,211	21.89
2021–22	8,939,518	103,586	34.77

Source Bangladesh Foreign Trade Statistics 2020–21

Table 5.5 Import by major commodities, 2017–18 to 2021–22

Year	Consumer goods	Materials for consumer goods	Capital goods	Materials for capital goods	Total
2017–18	888,037	3,716,874	844,755	61,978	5,511,644
2018–19	568,881	4,235,415	870,972	47,407	5,722,675
2019–20	540,221	4,160,831	691,419	49,187	5,441,658
2020–21	690,414	5,177,557	707,580	57,501	6,633,052
2021–22	1,008,356	7,039,391	830,670	61,101	8,939,518

Source Bangladesh Foreign Trade Statistics 2020–21

5.4 Major Export and Import Destinations of Bangladesh

Bangladesh, over the past few decades, has emerged as a significant player in the global trade arena, boasting a diverse export basket and an ever-expanding array of import sources. Predominantly known for its textile and garment industry, the nation has successfully positioned itself as a major apparel exporter to the developed world, especially to the European Union and the United States. These regions not only serve as substantial consumers of Bangladeshi goods but are also pivotal for technological and capital goods imports that fuel Bangladesh's industries. On the eastern front, countries like China and India play a multifaceted role in Bangladesh's trade narrative. China, as of recent years, has become one of the primary sources of imports, encompassing a range of goods from machinery to consumer products. Meanwhile, India serves as both a significant source of imports and a burgeoning export destination, underpinned by geographical proximity and shared cultural ties.

As the global economic landscape continues to evolve, understanding the dynamics of Bangladesh's primary trade partners is crucial for policymakers, businesses, and academic researchers keen on mapping the nation's external economic engagements and predicting future trajectories.

The data (Table 5.6) provides a snapshot of Bangladesh's import landscape over three consecutive fiscal years: 2019–20, 2020–21, and 2021–22. A few key observations emerge from the given statistics:

Sustained Growth in Imports: The total import value has been on a consistent upward trajectory, registering growth from 5,441,658 million Tk. in 2019–20 to a significantly higher 8,939,518 million Tk. by 2021–22.

Dominance of China and India: China and India, two major Asian economies, have consistently held the top two positions in terms of import values. China's contribution as a percentage of the total import has slightly increased, moving from 20.64% in 2019–20 to 21.33% in 2021–22. India, notably, has seen a more substantial growth in its percentage share, escalating from 12.07% in 2019–20 to 16.97% in 2021–22, indicating a deepening economic connection between the two neighbors.

Table 5.6 Imports from top ten and other countries, 2019–20, 2020–21, 2021–22 (million Tk.)

Country	2019–20		2020–21		2021–22	
	Value	% of total	Value	% of total	Value	% of total
Total import	5,441,658	100.00	6,633,052	100.00	8,939,518	100.00
1. China	1,123,512	20.64	1,350,427	20.36	1,907,188	21.33
2. India	656,981	12.07	1,059,725	15.98	1,516,786	16.97
3. Singapore	254,994	4.69	558,612	8.42	722,309	8.08
4. Indonesia	421,190	7.74	282,766	4.26	445,288	4.98
5. Hongkong	179,697	3.30	265,785	4.01	348,649	3.90
6. Brazil	128,969	2.37	235,063	3.54	342,436	3.83
7. Malaysia	53,425	0.98	231,221	3.49	293,504	3.28
8. Japan	216,361	3.98	204,027	3.08	275,695	3.08
9. USA	219,741	4.04	195,837	2.95	231,596	2.59
10. Korea	87,056	1.60	121,526	1.83	47,479	0.53
11. Others Countries	2,099,732	38.59	2,128,063	32.08	2,737,990	30.63
Total import	5,441,658	100.00	6,633,052	100.00	8,939,518	100.00

Source Bangladesh Foreign Trade Statistics 2020–21

Shifting Dynamics with Other Countries: Singapore, which ranked third, witnessed a significant increase in its share from 4.69% to 8.42% between 2019–20 and 2020–21 but tapered slightly in 2021–22. Contrarily, Indonesia saw a decline in its share in 2020–21 but rebounded in 2021–22. The USA's import share gradually decreased over the three years, while Korea experienced a sharp decline in its share, from 1.60% in 2019–20 to just 0.53% in 2021–22.

Contribution of 'Other' Countries: The collective import share from 'Other Countries' has decreased over the three years, from 38.59% in 2019–20 to 30.63% in 2021–22, suggesting a possible concentration of imports from major trading partners or a diversification in sourcing from other non-listed countries.

Overall Trade Diversification: While certain countries like China and India have strengthened their positions, the shifting dynamics with other nations indicate a dynamic and evolving trade strategy on Bangladesh's part, reflecting global economic conditions, bilateral relations, and domestic demand factors.

In essence, Bangladesh's import data over these years provides a lens into the country's economic dependencies, emerging trends, and strategic shifts in its global trade engagements. It is imperative for policymakers to monitor these patterns to ensure economic stability and capitalize on emerging trade opportunities (Table 5.7).

Table 5.7 Import payments under (cash + buyer's credit + IDB/ITFC + loans and grants) of Bangladesh with top twenty countries

		Taka in crore		In million US$	
Year		2021–2022	% of total	2021–2022	% of total
S. No.	Major countries				
1	China	167,149.3	25.6	19,349.5	25.6
2	India	118,287.1	18.1	13,690.0	18.1
3	Singapore	35,510.6	5.4	4115.8	5.4
4	Malaysia	30,091.0	4.6	3471.5	4.6
5	Indonesia	26,598.2	4.1	3081.7	4.1
6	USA	24,467.4	3.7	2825.7	3.7
7	Japan	21,021.0	3.2	2435.8	3.2
8	Brazil	19,426.5	3.0	2245.1	3.0
9	Qatar	18,904.1	2.9	2177.6	2.9
10	Saudi Arabia	14,644.9	2.2	1693.3	2.2
11	UAE	14,266.0	2.2	1646.8	2.2
12	Korea,	12,897.0	2.0	1492.6	2.0
13	Republic of Australia	10,731.9	1.6	1239.0	1.6
14	Taiwan: Province of China	9539.1	1.5	1101.8	1.5
15	Thailand	9093.4	1.4	1053.0	1.4
16	Vietnam	8726.5	1.3	1010.0	1.3
17	Germany	8317.4	1.3	962.8	1.3
18	Morocco	8001.2	1.2	919.3	1.2
19	Argentina	6843.7	1.0	791.5	1.0
20	Pakistan	5582.4	0.9	645.9	0.9
21	Other countries	83,370.9	12.8	9655.7	12.8
Total		653,469.6	100.0	75,604.4	100.0

Source Bangladesh Bank Import Report 2021–22

5.5 Direction of Exports by Country

Bangladesh exports more in terms of number of items and value to the United States of America and European Countries.

Table 5.8 presented offers a comprehensive view of the trajectory of Bangladesh's exports to its primary international partners over the span of three fiscal years, from 2019–20 to 2021–22. A deeper analysis of the data reveals several noteworthy trends and observations:

US Dominance in Exports: The United States emerges as the predominant export destination for Bangladesh, experiencing consistent growth both in value and its share of the total exports. Starting with a contribution of 16.30% in 2019–20, it rose to

Table 5.8 Countries where Bangladesh exports more (value in million Tk.)

Countries where exports	2019–20		2020–21		2021–22	
	Value	% of total	Value	% of total	Value	% of total
1. U.S.A	458,342	16.30	607,092	17.45	904,252	19.78
2. Germany	37,945	1.35	528,060	15.17	658,659	14.41
3. UK	255,589	9.09	328,914	9.45	417,709	9.14
4. Spain	129,727	4.61	214,777	6.17	273,310	5.98
5. France	82,202	2.92	173,013	4.97	235,024	5.14
6. Italy	533,943	18.99	119,667	3.44	177,757	3.89
7. India	98,799	3.51	114,558	3.29	147,532	3.23
8. Japan	162,189	5.78	104,366	3.00	132,312	2.89
9. Canada	54,510	1.94	102,844	2.96	117,552	2.57
10. Australia	31,854	1.13	71,749	2.06	79,065	1.73
11. Other countries	966,568	34.38	1,114,850	32.04	1,428,182	31.24
Total exports	2,811,668	100.00	3,479,890	100.00	4,571,354	100.00

Source Bangladesh Export Promotion Bureau, 2021–22

a significant 19.78% by 2021–22, underscoring the strengthening trade relationship between the two nations.

European Partners: Germany, the UK, Spain, France, and Italy collectively constitute a major portion of Bangladesh's export market. While Germany showed a marked increase in its share from 2019–20 to 2020–21, it experienced a slight decline in 2021–22. Surprisingly, Italy's contribution plummeted from 18.99% in 2019–20 to 3.89% in 2021–22, indicating possible trade shifts or challenges.

Asian Markets: India and Japan, representing two significant Asian economies, have maintained relatively stable but modest contributions to Bangladesh's exports. Both countries experienced a marginal decline in their respective shares over the three years.

Other Destinations: Countries like Canada and Australia demonstrated growth in terms of their shares in Bangladesh's exports, though their overall contribution remains below 3% of the total export value.

The export data underscores Bangladesh's increasing reliance on the US market and the importance of European countries in its export basket. The evolving dynamics with nations like Italy, coupled with the consistent contribution from 'Other countries', highlights the necessity for diversification and the potential to explore emerging markets to ensure sustainable economic growt (Table 5.9).

Table 5.9 Export receipts of Bangladesh from major countries

S. No.	Country/ commodities	2021–2022 1	2020–2021 2	2019–2020 3	Changes (1–2) 4	Changes (1–3) 5
1	U.S.A	7397	4946	3984	2451	3413
		(19.2)	*(16.5)*	*(15.1)*		
	Readymade garments	6897	4590	3721	2307	3176
	Fish, shrimps and prawns	40	36	18	4	22
	Home textiles	58	58	30	0	28
	Jute manufactures	18	21	15	− 3	3
	Leather and leather manufactures	105	63	41	42	64
	Raw jute	3	1	1	2	2
	Handicrafts	6	4	3	2	3
	Others	270	173	155	97	115
2	Germany	5081	4245	3893	836	1188
		(13.2)	*(14.2)*	*(14.8)*		
	Readymade garments	4724	3956	3639	768	1085
	Fish, shrimps, and prawns	58	37	46	21	12
	Home textiles	93	92	65	1	28
	Jute manufactures	6	5	2	1	4
	Leather and leather manufactures	73	60	72	13	1
	Raw jute	3	2	2	1	1
	Handicrafts	2	2	1	0	1
	Others	122	91	66	31	56
3	U.K	3543	2848	2690	695	853
		(9.2)	*(9.5)*	*(10.2)*		
	Readymade garments	3319	2636	2500	683	819
	Fish, shrimps, and prawns	64	62	71	2	− 7

(continued)

Table 5.9 (continued)

S. No.	Country/ commodities	2021–2022	2020–2021	2019–2020	Changes (1–2)	Changes (1–3)
		1	2	3	4	5
	Home textiles	61	63	52	– 2	9
	Jute manufactures	4	4	3	0	1
	Leather and leather manufactures	19	10	13	9	6
	Raw jute	3	3	3	0	0
	Handicrafts	2	2	1	0	1
	Others	71	68	47	3	24
4	Spain	2629	1993	1925	636	704
		(6.8)	*(6.6)*	*(7.3)*		
	Readymade garments	2527	1902	1838	625	689
	Fish, shrimps, and prawns	6	3	3	3	3
	Home textiles	13	10	9	3	4
	Jute manufactures	5	5	5	0	0
	Leather and leather manufactures	32	41	47	– 9	– 15
	Raw jute	1	1	1	0	0
	Handicrafts	4	2	1	2	3
	Others	41	29	21	12	20
5	France	2153	1688	1556	465	597
		(5.6)	*(5.6)*	*(5.9)*		
	Readymade garments	1977	1563	1463	414	514
	Fish, shrimps, and prawns	16	16	17	0	– 1
	Home textiles	54	51	37	3	17
	Jute manufactures	1	2	1	– 1	0

(continued)

Table 5.9 (continued)

S. No.	Country/ commodities	2021–2022	2020–2021	2019–2020	Changes (1–2)	Changes (1–3)
		1	2	3	4	5
	Leather and leather manufactures	71	31	18	40	53
	Raw jute	0	0	0	0	0
	Handicrafts	1	1	1	0	0
	Others	33	24	19	9	14
6	Poland	1703	1211	1027	492	676
		(4.4)	*(4.0)*	*(3.9)*		
	Readymade garments	1580	1124	961	456	619
	Fish, shrimps, and prawns	0	0	0	0	0
	Home textiles	34	18	11	16	23
	Jute manufactures	3	3	2	0	1
	Leather and leather manufactures	55	40	39	15	16
	Raw jute	0	0	0	0	0
	Handicrafts	3	1	0	2	3
	Others	28	25	14	3	14
7	India	1504	1008	866	496	638
		(3.9)	*(3.4)*	*(3.3)*		
	Readymade garments	423	238	258	185	165
	Fish, shrimps, and prawns	62	76	32	– 14	30
	Home textiles	2	0	0	2	2
	Jute manufactures	111	106	124	5	– 13
	Leather and leather manufactures	46	25	15	21	31
	Raw jute	71	30	44	41	27
	Handicrafts	0	0	0	0	0

(continued)

Table 5.9 (continued)

S. No.	Country/ commodities	2021–2022	2020–2021	2019–2020	Changes (1–2)	Changes (1–3)
		1	2	3	4	5
	Others	789	533	393	256	396
8	Italy	1427	1210	1206	217	221
		(3.7)	*(4.0)*	*(4.6)*		
	Readymade garments	1351	1145	1139	206	212
	Fish, shrimps, and prawns	1	2	1	− 1	0
	Home textiles	14	14	11	0	3
	Jute manufactures	3	2	1	1	2
	Leather and leather manufactures	25	22	33	3	− 8
	Raw jute	0	0	0	0	0
	Handicrafts	1	1	0	0	1
	Others	32	24	21	8	11
9	Netherlands	1322	965	832	357	490
		(3.4)	*(3.2)*	*(3.2)*		
	Readymade garments	1104	794	662	310	442
	Fish, shrimps, and prawns	77	56	70	21	7
	Home textiles	19	11	7	8	12
	Jute manufactures	16	16	13	0	3
	Leather and leather manufactures	46	36	47	10	− 1
	Raw jute	2	1	0	1	2
	Handicrafts	7	4	2	3	5
	Others	51	47	31	4	20
10	Canada	1107	870	772	237	335
		(2.9)	*(2.9)*	*(2.9)*		
	Readymade garments	1032	790	717	242	315

(continued)

Table 5.9 (continued)

S. No.	Country/ commodities	2021–2022	2020–2021	2019–2020	Changes (1–2)	Changes (1–3)
		1	2	3	4	5
	Fish, shrimps, and prawns	4	3	2	1	2
	Home textiles	34	41	28	− 7	6
	Jute manufactures	1	2	1	− 1	0
	Leather and leather manufactures	9	4	4	5	5

Source Bangladesh Foreign Trade Statistics, 2021–22

5.6 Major Products for Export and Import of Bangladesh

Bangladesh, sometimes referred to as the 'Golden Fiber' of the East because of its long history of producing jute, has seen a substantial change in its trading structure in recent years. The export–import dynamics of the country now provide an intriguing window into the changing industrial and economic might of a fast-rising South Asian nation. The ready-made garment (RMG) industry, which accounts for a sizeable share of the nation's overall exports and has come to symbolize the export identity of the nation, is at the center of this shift. Products including knitwear, woven clothing, and specialty textiles have entered important global markets, solidifying Bangladesh's standing as one of the top exporters of apparel worldwide.

Bangladesh has a wide range of import requirements, which are a reflection of its expanding consumer market and needs for development. The country's efforts to support its industrial base and meet its energy requirements have resulted in a large portion of the import bill being made up of petroleum and petroleum products, as well as machinery for industry. Additionally, to fill the gap between domestic production and consumption, agricultural items such as grains, raw cotton, and culinary oils are imported. The import of raw materials, such as fabrics and accessories, for the RMG industry is another significant factor. These materials are added to value domestically before being exported as completed goods. For a thorough examination of Bangladesh's trade dynamics, it is crucial to comprehend the main products in the country's export and import basket. These goods highlight the nation's economic priorities as well as its potential and problems within the global trade environment. It is critical to keep an eye on these essential goods and how they influence Bangladesh's trade story as the country grows and diversifies.

5.6.1 Imports of Principal Goods During Last Five Years

According to Table 5.10 Over the last half-decade, Bangladesh's import trends for principal goods highlight the nation's evolving consumption patterns, industrial needs, and responses to external market dynamics. Here's an analytical breakdown of the data:

Agricultural Commodities: The imports of basic food items such as milk, sugar, rice, and wheat have seen fluctuations. Notably, sugar and wheat have witnessed a significant increase over the years, pointing towards either growing domestic consumption or potential shortfalls in local production. Interestingly, rice, a staple in the Bangladeshi diet, experienced a marked decline in imports by 2021–22, perhaps signifying successful local agricultural policies or bumper harvests.

Industrial Inputs: The trends in raw cotton, petroleum products, and textile yarn underscore the importance of the textile and energy sectors. While raw cotton saw a drastic uptick in 2021–22, textile yarn imports dwindled. This could imply a shift in the local textile industry's sourcing dynamics or changes in global cotton prices. Meanwhile, the reduction in crude oil imports might be due to diversifying energy sources or improvements in domestic refining capacities.

Infrastructure and Development: The steady rise in iron & steel imports points towards extensive infrastructural development, aligning with Bangladesh's accelerated urbanization and construction boom. Simultaneously, the significant drop in

Table 5.10 The imports of principal goods during the last five years (million Tk.)

Principal goods	2016–17	2017–18	2018–19	2019–20	2020–21	2021–22
Milk and cream	24,831	28,639	31,900	28,559	32,439	40,228
Sugar	90,737	78,171	70,714	76,482	131,403	190,335
Rice	10,238	136,048	10,113	136,048	75,552	37,971
Wheat	107,186	124,791	12,363	124,790	162,929	193,775
Raw cotton	1888	272,378	31,330	272,378	323,410	397,441
Petroleum oil (crude)	40,356	29,925	34,881	29,925	221,808	80,793
Petroleum products	318,497	502,620	331,251	354,453	328,176	311,753
Fertilizer	66,485	90,271	11,919	90,272	12,965	416,373
Plastics and articles	172,140	191,460	220,707	222,588	266,645	351,309
Edible oil	273,069	283,464	246,829	283,464	74,569	453,533
Iron & steel	202,003	234,569	276,335	295,029	276,079	472,225
Machinery and equipment inclu. Elec. Appliances	754,457	979,686	967,987	755,753	420,388	121,790
Textile yarn (cotton & synthetic)	129,332	150,738	88,877	150,738	76,778	27,814

Source Bangladesh Foreign Trade Statistics 2020–21

machinery and equipment imports in 2021–22 could be a result of completed major projects or a move towards local manufacturing.

Consumer Goods and Raw Materials: Imports of plastics and edible oil have risen considerably, possibly indicating a growing consumer market and the reliance on these essential raw materials. The spike in edible oil imports in 2021–22 might be due to increased global prices, local production gaps, or strategic stockpiling.

Special Mentions: The explosive growth in fertilizer imports in 2021–22 is remarkable. Such a surge could be attributed to agricultural policy changes, increased farming activities, or a response to prior shortages.

Bangladesh's import trends offer a lens into its developmental trajectory, emphasizing the balance between self-reliance and the necessities of a growing economy. The data portrays a nation adept at adjusting to global market realities while striving for domestic production capabilities.

5.6.2 Unit Price of Selected Import Commodities

Import price is an important indicator for fixing up tariff and taxation, policy formulation and domestic market analysis. Table 5.11 shows the unit price of selected major commodities of imported goods during 2019–20 & 2020–21 (Table 5.12).

Table 5.11 Unit price of selected commodities, 2019–20 and 2020–21 (value in Tk.)

Selected commodities	Unit	2019–20	2020–21	% Change over 2020–21	2021–22	% Change over 2020–22
1. Rice	M.T	35,433	36,308	2.47	37,442	3.12
2. Wheat	M.T	18,029	24,086	33.60	31,115	29.18
3. Edible oil	M.T	59,566	74,569	25.19	114,183	53.12
4. Petroleum oil (crude)	M.T	24,757	24,239	– 2.09	25,537	5.36
5. Petroleum products	M.T	40,703	41,584	2.16	44,559	7.15
6. Plastics and articles	M.T	116,932	66,760	– 42.91	151,790	127.37
7. Raw cotton	M.T	163,895	175,603	7.14	214,102	21.92
8. Cotton yarn and synthetic yarn	M.T	271,697	252,113	– 7.21	743,849	195.05
9. Cement(Clinker)	M.T	423,034	424,001	0.23	5307	– 98.75
10. Man-made staple fiber	M.T	320,867	173,677	– 45.87	349,184	101.05
11. Fertilizers	M.T	25,734	29,339	14.01	60,501	106.21
12. Iron and steel	M.T	38,335	40,211	4.89	56,880	41.45

Source Bangladesh Foreign Trade Statistics 2020–21

Table 5.12 Major import commodities (In Million USD)

Major commodities	2021–2022		2020–2021		Changes (1)–(3)
	Amount	% of Total (A)	Amount	% of Total (A)	
	1	2	3	4	5
Import under (cash + buyer's credit + IDB/ ITFC loan + loans and grants)	75,604.4	100.0	50,987.6	100.0	+ 24,616.8
Mineral fuels, mineral oils, and products of their distillation bituminous substances; mineral, waxes	12,767.2	16.9	6100.1	12.0	+ 6667.0
Cotton	10,058.6	13.3	6326.7	12.4	+ 3731.9
Nuclear reactors, boilers, machinery, and mechanical appliances; parts thereof	7035.4	9.3	4504.2	8.8	+ 2531.1
Iron and steel	4271.6	5.6	2763.1	5.4	+ 1508.5
Plastics and articles thereof	3281.3	4.3	2641.6	5.2	+ 639.8
Cereals	3183.0	4.2	2399.3	4.7	+ 783.7
Electrical machinery and equipment and parts thereof; sound recorders and reproducers, television image and sound recorders and reproducers and parts and accessories of such articles	2890.9	3.8	2471.8	4.8	+ 419.1
Fertilizers	2891.4	3.8	974.0	1.9	+ 1917.5

(continued)

Table 5.12 (continued)

Major commodities	2021–2022		2020–2021		Changes (1)–(3)
	Amount	% of Total (A)	Amount	% of Total (A)	
	1	2	3	4	5
Animal or vegetable fats and oils and their cleavage products; prepared edible fats; animal or vegetable waxes	2559.4	3.4	1817.4	3.6	+ 742.0
Oil seeds and oleaginous fruits; miscellaneous grains; seeds and fruit; industrial or medicinal plants; straw and fodder	1682.6	2.2	1375.4	2.7	+ 307.2
Ships, boats, and floating structures	1662.9	2.2	1175.7	2.3	+ 487.1
Man-made staple fibers	1655.2	2.2	1072.1	2.1	+ 583.1
Man-made filaments	1656.3	2.2	1088.4	2.1	+ 567.9
Vehicles other than railway or tramway rollingstock, and parts and accessories thereof	1592.9	2.1	1336.2	2.6	+ 256.6
Knitted or crocheted fabrics	1581.3	2.1	990.6	1.9	+ 590.8

Source Bangladesh Foreign Trade Statistics 2020–21

Table 5.13 data presented offers valuable insights into the export trends of Bangladesh over a five-year span, emphasizing the strengths and changing dynamics of its export-oriented industries. Here's a comprehensive breakdown:

Readymade Garments: Undoubtedly, the backbone of Bangladesh's exports, the ready-made garment (RMG) sector, shows robust growth over the years, culminating in a significant jump in 2021–22. This likely reflects the sector's resilience in recovering from global market disruptions, such as the pandemic's impact in 2019–20, and further strengthening its global position.

Table 5.13 Exports of principal commodities during last five years (million Tk.)

Principal commodities	2017–18	2018–19	2019–20	2020–21	2021–22
Readymade garments	2,513,449	2,867,433	2,274,744	2,778,659	3,681,754
Made up textile articles	82,085	78,501	65,458	109,676	150,626
Vegetable textile fiber/yarn	74,119	61,665	59,228	89,865	87,044
Shrimps and prawn	35,094	34,244	35,094	29,372	36,187
Footwear	66,367	73,819	61,719	79,730	103,942
Hides, skins and leather	15,118	13,961	7978	14,723	13,145
Raw jute	12,781	9451	12,781	12,626	18,656
Special Woven fabrics	4037	4759	3288	3789	5557
Plastics and Articles Thereof	220,707	10,368	8355	10,325	14,483
Hats and other headgear	15,075	17,358	14,859	19,712	31,406

Source Bangladesh Foreign Trade Statistics 2020–21

Textile and Related Exports: While the RMG sector thrived, made-up textile articles and vegetable textile fiber/yarn also witnessed growth, albeit with some fluctuations. The consistent export figures suggest that Bangladesh has maintained a significant market share in these specialized textile categories.

Aquaculture Products: The relatively stable figures for shrimps and prawn exports denote a sustained demand in international markets, with a notable slight uptick in 2021–22, hinting at an industry rebound or expanded market reach.

Footwear: A promising upward trajectory is evident in the footwear sector, marking a diversification in Bangladesh's export portfolio. The consistent growth, especially the surge in 2021–22, underscores the country's emerging status as a significant global footwear producer.

Leather and Related Goods: The hides, skins, and leather segment, however, seems to be grappling with challenges given the declining trends, necessitating potential introspection for industry stakeholders.

Raw Jute: Despite some fluctuations, raw jute exports experienced growth in 2021–22, reflecting possible renewed global interest in eco-friendly, natural fibers.

Diverse Commodities: The export figures for special woven fabrics, plastics and articles, and hats indicate diverse capabilities within the Bangladeshi export framework. The particularly notable growth in hat exports in the latter years deserves special attention for potential expansion.

In summation, Bangladesh's export dynamics over the last five years spotlight the nation's prowess in the RMG sector, its expanding footprint in footwear and diverse commodities, and areas like leather that may need rejuvenation strategies. The data reinforces Bangladesh's significance in the global export landscape, highlighting areas of strength and potential opportunities (Tables 5.14 and 5.15).

Table 5.14 Unit price of principal export items during last six years (Value in Tk.)

Item	Unit	2016–17	2017–18	2018–19	2019–20	2020–21	2021–22
Tea	kg	158	264	361	264	192	291
Terry toweling	kg	210	231	235	231	236	292
Fertilizer	MT	14,240	–	–	–	690,961	–
Raw hides and skin	MT	1,246,489	1,314,485	940,530	1,314,600	604,835	495,676
Shrimps and prawns	kg	900	923	782	923	855	1030
Raw jute	MT	52,800	48,696	52,922	48,696	69,998	88,521
Jute goods	MT	84,292	87,313	81,998	87,313	117,048	133,828

Source Bangladesh Foreign Trade Statistics 2020–21

Table 5.15 Export receipts of Bangladesh by major commodities (Taka in crore)

Year/ Commodities	2021–2022	% of Total	2020–2021	% of Total	2019–2020	% of Total
Knit wear	171,772.7	51.6	127,037.2	49.9	107,433.1	48.1
Woven garments	110,866.4	33.3	85,062.0	33.4	83,441.6	37.3
Jute and jute manufactures	8762.2	2.6	9146.8	3.6	7186.1	3.2
Leather and leather manufactures	6342.2	1.9	4618.3	1.8	4398.1	2.0
Fish, shrimps, and prawns	4013.0	1.2	3485.2	1.4	3442.8	1.5
Home textile	5907.8	1.8	4336.1	1.7	3097.8	1.4
Vegetable products	1687.8	0.5	1815.8	0.7	1381.4	0.6
Pharmaceutical products	1161.2	0.3	1082.3	0.4	879.6	0.4
Plastic and plastic products	950.3	0.3	733.5	0.3	537.8	0.2
Bicycle	1024.6	0.3	862.5	0.3	450.9	0.2
Petroleum and petroleum products	7.1	0.0	93.7	0.0	9.1	0.0
Others	20,213.9	6.1	16,092.3	6.3%	11,290.1	5.1
Total	332,709.2	100.0	254,365.7	100.0	223,548.4	100.0

Source Bangladesh Foreign Trade Statistics 2020–21

In assessing the trade landscape of Bangladesh, it becomes unequivocally clear that the nation has etched a distinct niche in the global market, particularly with its export products. The ready-made garment sector stands out as a linchpin of the nation's exports, not only showcasing impressive numbers but also reflecting the resilience and adaptability of the industry in the face of global challenges. On the import front, the diverse range of principal goods, from essential commodities like edible oil and wheat to industrial requisites such as machinery and raw cotton, underlines the country's intertwined economic dependencies and growth aspirations.

It's pivotal to recognize that while the export basket, dominated by textiles, underpins Bangladesh's economic identity, the imports paint a picture of a nation gearing for industrial modernization and addressing its populace's basic needs. In conclusion, the major products for export and import of Bangladesh provide a nuanced narrative of a nation that is seamlessly integrating with the global economy while staying rooted in its core competencies, navigating its growth trajectory with foresight and ambition.

5.7 Current State of Bilateral Trade Between Bangladesh and LAC

In the ever-evolving tapestry of global trade dynamics, the bilateral relationships between nations, often distant in geography but close in strategic interests, hold the potential to reshape regional economies. One such relationship, burgeoning with promise yet underexplored, is between Bangladesh and the countries of Latin America. Bangladesh, recognized predominantly for its robust textile industry and increasing diversification of its export portfolio, has embarked on a journey to explore markets beyond its traditional trading partners. Latin American countries, with their vast natural resources, emergent manufacturing sectors, and a shared history of colonialism and subsequent development challenges, present a unique tapestry of opportunities and synergies for Bangladesh.

The current state of bilateral trade between Bangladesh and Latin American countries, while still in its nascent stages, has shown encouraging signs. Various studies indicate that there exists a considerable untapped potential for enhancing trade ties, with both sides displaying keen interest in expanding their economic engagements. Key commodities like jute and textiles from Bangladesh find a burgeoning market in Latin America, while Bangladesh, in turn, has shown interest in Latin American agricultural products, minerals, and raw materials. Yet, for all its promise, there are hurdles to be addressed. Infrastructural challenges, lack of direct shipping routes, and limited awareness of market needs on both sides have occasionally stymied growth. Moreover, while both regions are keen to diversify their trading portfolios and reduce dependency on traditional markets, there exists a knowledge gap about each other's capabilities and strengths. However, with the increasing number of bilateral meetings, trade delegations, and collaborative platforms being set up, the trajectory is certainly upward. As both Bangladesh and Latin America pivot towards a more inclusive and diversified global trade strategy, understanding the currents of this bilateral relationship becomes pivotal for trade analysts, policymakers, and stakeholders.

5.7.1 *Bangladesh's International Trade with Latin American Countries (2021–22): An Analysis*

Bangladesh's international trade with South American countries reveals a striking dynamic (Table 5.16). In the fiscal year 2021–22, the country exported goods worth approximately TK 44.56 billion to the region while importing significantly more, about TK 430.98 billion. Delving deeper into country-specific data illuminates the trade landscape. Brazil stands out as Bangladesh's dominant trading partner in the region, accounting for a substantial portion of both exports (TK 9.44 billion) and, more significantly, imports (TK 275.69 billion). This stark trade imbalance underscores Brazil's salient role in Bangladesh's import structure.

Chile, surprisingly, emerges as a major export destination for Bangladesh, with exports valued at TK 16.56 billion, juxtaposed against a relatively meager import of TK 1.53 billion. Similarly, significant export figures are observed for countries

Table 5.16 Bangladesh's international trade with Latin American countries (2021–22)

Country by area	2021–22	
	Export	Import
South America	44,563,857,714	430,981,998,381
Argentina	821,156,421	131,399,734,079
Bolivia	276,544,757	1,878,925
Brazil	9,441,802,286	275,695,062,289
Chile	16,565,394,109	1,531,103,003
Colombia	5,988,093,244	99,263,855
Ecuador	399,067,306	75,959,579
Falkland ISL	0	2614
French Guiana	4,267,519	434
Guyana	16,795,575	39,571,086
Paraguay	516,258,121	12,574,679,260
Peru	6,370,062,634	3,587,991,684
Surinam	7,884,998	491,405,145
Uruguay	3,350,960,444	4,254,953,983
Venezuela	805,570,300	1,230,392,445

Source Bangladesh Foreign Trade Statistics 2021–22 and National Board of Revenue

like Colombia (TK 5.98 billion) and Peru (TK 6.37 billion). However, for some nations, the trade volume is minuscule or even non-existent. The Falkland Islands, for instance, register negligible trade activity with no recorded exports and an inconsequential import of TK 2614. French Guiana too mirrors this pattern with marginal trade figures. In essence, while Bangladesh has a diversified trading relationship with several South American nations, evident disparities exist. Some countries emerge as pivotal trade partners, while others barely feature in the trade narrative. The data suggests potential areas for trade expansion and diversification for Bangladesh in the region.

5.7.2 Analysis of Bangladesh's International Trade with Latin American Countries (2020–21)

The 2020–21 fiscal year witnessed Bangladesh's burgeoning trade interactions with South America, (Table 5.17) amassing exports worth approximately TK 27.94 billion and imports valued significantly higher at TK 360.38 billion. A deeper scrutiny of the country-specific data reveals notable patterns. Brazil, undisputedly, remains Bangladesh's principal trading partner within the region, registering exports of TK 8.41 billion juxtaposed against a staggering TK 235.06 billion in imports. This evident trade deficit accentuates Brazil's dominant influence in Bangladesh's import dynamics. Interestingly, Chile emerges prominently in Bangladesh's export landscape with exports amounting to TK 8.09 billion, while its imports remained relatively modest at TK 2.87 billion. Peru and Colombia also feature as notable export destinations with values of TK 4.28 billion and TK 3.18 billion, respectively.

Contrastingly, trade volumes with countries like Bolivia, French Guiana, and the Falkland Islands remain insubstantial, with the latter two recording no trade activity for the fiscal year. In summation, Bangladesh's trade relationship with South American countries during 2020–21 presents a mixed tableau. While interactions with certain nations like Brazil and Chile are pronounced, others remain peripheral, indicating potential avenues for trade enhancement and diversification in future.

5.7.3 Analysis of Bangladesh's International Trade with Latin American Countries (2019–20)

During the fiscal year 2019–20, (Table 5.18) Bangladesh's trade trajectory with South America painted a dynamic picture, clocking exports of roughly TK 29.16 billion vis-à-vis imports of about TK 301.13 billion. A nuanced examination of the figures unfolds distinctive trends. Brazil prominently surfaces as Bangladesh's primary trading ally within South America, with exports registering at $8.82 billion and imports at an even more substantial TK 197.25 billion. This highlights a pronounced

Table 5.17 Bangladesh's international trade with Latin American countries (2020–21)

Country by area	2020–21	
	Export	Import
South America	27,937,057,434	360,382,719,431
Argentina	664,055,679	110,701,439,824
Bolivia	331,868,007	0
Brazil	8,414,074,610	235,063,186,911
Chile	8,093,632,732	2,868,858,813
Colombia	3,180,428,242	126,140,099
Ecuador	199,089,520	33,367,948
Falkland ISL	0	7375
French guiana	0	0
Guyana	35,841,406	45,688,805
Paraguay	299,307,583	4,279,127,126
Peru	4,280,762,698	2,052,777,302
Surinam	13,927,054	1,229,645,028
Uruguay	1,827,328,575	3,877,126,486
Venezuela	596,741,328	105,353,714

Source Bangladesh Foreign Trade Statistics 2020–21 and National Board of Revenue

trade imbalance skewed towards imports from Brazil. Chile also figures dominantly, with exports amounting to TK 8.11 billion, while imports were comparatively modest at TK 5.08 billion. Notably, countries like Peru, Colombia, and Uruguay emerge as significant contributors to Bangladesh's export canvas, each accounting for billions in trade.

Conversely, trade interactions with countries such as Bolivia, French Guiana, and the Falkland Islands remain on the lower end, with the latter showcasing negligible trade activity. In essence, the 2019–20 fiscal year delineated Bangladesh's intense trade engagements with certain Latin American nations, while relations with others were relatively subdued. These patterns underscore potential regions for further trade exploration and diversification for Bangladesh in the coming years.

5.7.4 *Comparison of Bangladesh's International Trade with Latin American Countries (2019–20 to 2021–22)*

Over the three fiscal years from 2019–20 to 2021–22, Bangladesh's trade with Latin American countries has witnessed notable fluctuations and shifts. A consolidated analysis of the data provided unfolds a series of trends and comparative insights.

Table 5.18 Bangladesh's international trade with Latin American countries (2019–20)

Country by area	2019–20	
	Export	Import
South America	29,157,046,379	301,129,342,713
Argentina	1,063,891,727	88,343,160,675
Bolivia	51,603,008	–
Brazil	8,821,465,708	197,247,892,664
Chile	8,115,142,564	5,084,214,270
Colombia	2,860,308,225	141,792,645
Ecuador	121,186,809	44,324,398
Falkland ISL	–	156,883
French Guiana	3,224,532	436
Guyana	22,211,223	32,641,753
Paraguay	87,547,646	2,244,280,951
Peru	4,659,852,065	3,795,534,926
Surinam	12,737,075	1,384,181,183
Uruguay	3,129,670,509	2,345,253,168
Venezuela	207,935,288	465,908,761

Source Bangladesh Foreign Trade Statistics 2019–20 and National Board of Revenue

Overall Trade Dynamics: From 2019–20 to 2021–22, Bangladesh's exports to South America surged from TK 29.16 billion to TK 44.56 billion, marking an impressive growth of over 50%. Simultaneously, imports escalated from TK 301.13 billion in 2019–20 to TK 430.98 billion in 2021–22, indicating an increased dependence on the region (Table 5.19).

Country-Specific Analysis: Brazil consistently emerges as the predominant trading partner. Exports to Brazil have grown, reaching TK 9.44 billion in 2021–22 from TK 8.82 billion in 2019–20. More prominently, imports have skyrocketed, moving from TK 197.25 billion in 2019–20 to TK 275.69 billion in 2021–22. This underscores Brazil's intensifying role in Bangladesh's import matrix.

Chile's significance in Bangladesh's export portfolio is evident. Exports have remained stable, oscillating around the TK 8 billion mark across the three years. Argentina's role as an import partner has surged, with imports almost doubling from TK 88.34 billion in 2019–20 to TK 131.40 billion in 2021–22. On the other hand, trade with countries like Bolivia, French Guiana, and the Falkland Islands remains subdued. Remarkably, Bolivia's exports surged to TK 276.54 million in 2021–22, after having a marginal presence in the previous years.

Key Takeaways: Over the span of these three fiscal years, Bangladesh has expanded its trade footprint in South America. The consistent rise in both exports and imports testifies to the deepening economic ties between Bangladesh and the Latin

Table 5.19 Bangladesh exports to LAC compare to other regions and trade blocs

Countries	2018–19	2019–20	2020–21	2021–22	2022–23
African	320.63	276.8	396.94	447.76	399.94
American	8922.82	7414.49	8655.15	12,848.8	12,534.09
ASEAN	698.9	585.76	714.83	841.95	923.88
Asian	4957.02	4212.24	4720.33	5938.13	6765.01
BIMSTEC	1415.83	1248.91	1472.95	2247.51	2369.52
CIS	605.68	545.02	733.31	699.5	529.23
D-8	853.61	843.11	1035.68	1086.53	1076.48
East and Other European	1004.03	987.74	1218.31	1158.78	1062.06
EU	22,855.34	18,698.25	17,464.35	23,231.71	25,235.02
Eurasian	558.89	498.66	677.05	645.93	470.83
EUROZONE	15,416.63	12,590.82	14,075.83	18,569.89	20,344.91
IORA	3180.85	2702.05	3411.8	4679.49	4935.78
LDC	175.87	168.86	252.28	276.99	221.60
Middle East	813.21	750.94	987.76	1386.71	1057.69
North America	8474.57	7048.61	8352.94	12,332.83	11,917.35
Ocenia	903.15	766.5	949.89	1047.67	1407.12
OIC	1725.36	1645.19	2128.76	2563.98	2232.27
SAARC	1408.27	1246.59	1499.92	2281.33	2362.15
South America	448.09	365.97	302.07	515.5	621.27

Source Export Promotion Bureau (2023)

American region. Certain countries, especially Brazil, have consolidated their positions as vital trade partners, while others offer potential avenues for future exploration and collaboration.

5.7.5 *Analysis of Bangladesh's Total Import Payments to LAC (2020–2021 and 2021–2022)*

Table 5.20 represents Bangladesh's import payments to selected Latin American countries (LAC) over the fiscal years 2020–2021 and 2021–2022, denominated in 'Taka in Lac'.

Argentina: For the fiscal year 2020–2021, the total import payments to Argentina stood at 529,043 Lac Taka, all under 'Cash + Buyer's Credit'. The subsequent fiscal year witnessed an increase of approximately 29.4%, with the figure rising to 684,375 Lac Taka.

Table 5.20 Total import payments of Bangladesh to LAC: Taka in Lac

Country	2020–2021					2021–2022				
	Cash + buyer's credit	Loans and grants	IDB loan (short-term)	Other un-classified imports	Total	Cash + buyer's credit	Loans and grants	IDB loan (short-term)	Other un-classified imports	Total
Argentina	529,043				529,043	684,375				684,375
Brazil	1,474,095				1,474,095	1,942,648				1,942,648
Paraguay	39,626				39,626	120,733				120,733
Peru	20,459				20,459	31,464				31,464
Uruguay	30,087				30,087	41,481				41,481

Source Bangladesh Bank 2021–22

Brazil: Brazil's dominance as a significant trade partner is evident. In 2020–2021, Bangladesh's import payments to Brazil amounted to 1,474,095 Lac Taka, with the entire amount allocated under 'Cash + Buyer's Credit'. By 2021–2022, the figure saw a substantial increment of 31.8%, reaching 1,942,648 Lac Taka.

Paraguay: Although Paraguay's import figures are relatively modest compared to Argentina and Brazil, there was a remarkable growth between the two fiscal years. From 39,626 Lac Taka in 2020–2021, the import payments jumped by 204.3% to 120,733 Lac Taka in 2021–2022.

Peru: The import payments to Peru in 2020–2021 were 20,459 Lac Taka. The subsequent year witnessed an increment of 53.8%, taking the figure to 31,464 Lac Taka in 2021–2022.

Uruguay: For Uruguay, the import payments in 2020–2021 were 30,087 Lac Taka. This escalated by 37.9% to 41,481 Lac Taka in 2021–2022.

In summary, the data highlights Bangladesh's deepening trade relationships with these Latin American nations. Every country on the list saw an increase in import payments from Bangladesh in the 2021–2022 fiscal year compared to the previous one. This uptrend underscores the growing reliance and economic ties between Bangladesh and the LAC region (Table 5.21).

5.7.6 Analysis of Bangladesh's Total Export Receipts to Latin American Countries (2005–2013)

Between 2005 and 2013, Bangladesh's export receipts from several Latin American countries reveal intriguing patterns and growth trajectories. Argentina showed a rising trend in export receipts until the 2011–2012 fiscal year, post which there was a decline in 2012–2013. Starting with 69,455 (in thousand Taka) in 2005–2006, the peak was reached in 2011–2012 at 360,038, a more than fivefold increase over the span. For Bolivia, export receipts experienced fluctuations throughout the period, with the highest being in 2007–2008 at 27,149 and subsequently observing a relative increase in 2012–2013 at 16,337. Brazil, a major trading partner, recorded a significant upswing in receipts. From 413,827 in 2005–2006, the numbers soared, almost tripling by 2008–2009, and reaching its peak at 11,768,090 in 2012–2013. Chile and Colombia consistently saw rising trends, with the former almost quadrupling its receipts from 2005 to 2012–2013 and the latter more than doubling. Ecuador and Guyana witnessed irregular patterns, with highs and lows throughout the period. However, Paraguay and Surinam observed noticeable peaks in the middle years before declining in 2012–2013. Peru's receipts trend was somewhat oscillatory but remained largely positive, reaching 460,966 in 2012–2013. Uruguay had a significant spike in 2009–2010 but saw a decline in the subsequent years. Venezuela saw a surge up to 2011–2012 and a slight dip in 2012–2013, while the Republic of Guinea fluctuated throughout, peaking in 2009–2010. In summary, the majority of Latin American

Table 5.21 Bangladesh's past records of total export receipts of Latin American countries: (Taka in Thousand)

Country	2005–2006	2006–2007	2007–2008	2008–2009	2009–2010	2010–2011	2011–2012	2012–2013
Argentina	69,455	107,395	161,536	272,981	169,286	315,365	360,038	314,096
Bolivia	8646	904	27,149	9201	9026	3667	6597	16,337
Brazil	413,827	711,230	1,341,354	3,793,181	3,942,277	5,414,811	9,857,942	11,768,090
Chile	53,740	55,508	213,449	384,625	391,231	781,640	1,184,801	1,739,763
Colombia	27,576	163,124	91,137	176,092	57,721	283,579	558,096	701,496
Ecuador	8636	72,739	482	16,239	5222	353	9972	51,796
Paraguay	2175	7201	5549	12,715	10,663	11,465	57,564	36,789
Peru	249,129	171,740	344,078	321,976	125,111	314,821	409,007	460,966
Surinam	0	2011	11,498	266	3040	7667	4410	18,702
Uruguay	34,259	91,077	127,383	116,264	329,630	267,783	330,415	292,085
Venezuela	23,443	23,732	36,606	66,727	62,748	80,591	187,874	176,289
Guinea Republic of	17,762	21,450	41,304	11,932	52,142	41,216	31,723	39,094
Guyana	299	0	4732	703	1155	0	6384	5572

Source Bangladesh Bank, Annual Report 2020

nations showed an increasing interest in Bangladeshi exports over the years, with Brazil leading in terms of value. This underscores Bangladesh's growing economic outreach and the potential for expanded trade ties with the Latin American region.

5.8 Potential Exports Products of Bangladesh Towards LAC: Case of Argentina and Brazil

Bangladesh, a South Asian nation, has showcased impressive economic growth in recent decades, partly attributable to its robust export-oriented industries. With Latin American Countries (LAC) progressively becoming influential in the global trade landscape, there lies an opportunity for Bangladesh to diversify its export destinations. This paper examines the potential export products Bangladesh can offer to two of the most significant LAC economies, Argentina and Brazil.

Textile and Ready-Made Garments (RMG)

Bangladesh's textile industry, especially RMG, stands as the linchpin of its export economy. With Argentina and Brazil's rising middle class and demand for affordable clothing, Bangladesh's RMG sector could find a lucrative market in these countries. Leveraging its competitive manufacturing costs and improving its supply chain logistics could be pivotal in penetrating these markets.

Jute and Jute Products

Renowned as the 'Golden Fiber' of Bangladesh, jute holds historical importance for the nation's economy. As global trends lean towards environmentally sustainable products, the demand for jute, a natural and biodegradable material, is on the rise. Both Argentina and Brazil, given their agricultural orientation, could benefit from importing jute products, including sacks for agricultural goods storage and transportation.

Leather and Leather Goods

Bangladesh's leather industry, noted for its quality and competitive pricing, could cater to the fashion and utility needs of Argentina and Brazil. With appropriate marketing strategies highlighting quality and affordability, Bangladeshi leather products like shoes, belts, and bags could gain traction in these LAC nations.

Seafood

The fisheries sector in Bangladesh, abundant in items like shrimp and Hilsa fish, has an established export market in Europe and the USA. Argentina, with its vast coastline, and Brazil, with its rich culinary culture, could be potential markets for Bangladeshi seafood, provided there are adjustments to meet specific tastes and preferences.

Pharmaceuticals

Bangladesh's burgeoning pharmaceutical sector, known for producing high-quality generic drugs at competitive prices, has export potential in LAC countries, including Argentina and Brazil. With increasing healthcare demands, these nations could benefit from affordable medicines and healthcare products from Bangladesh.

Information and Communication Technology (ICT) and Software Services

The ICT sector of Bangladesh has witnessed a surge in recent years. As Brazil and Argentina are fast-embracing digital solutions, there's potential for Bangladeshi IT firms to offer software solutions, mobile applications, and BPO services tailored for these markets.

Agro-Processed Foods

Bangladesh has a rich agricultural base, producing a variety of tropical fruits, spices, and grains. Agro-processed products such as jams, pickles, dried fruits, and spice mixes might find a consumer base in the Latin American markets, given the region's affinity for varied cuisines.

Ceramics and Tableware

Bangladesh's ceramic industry, known for its artistic designs and quality, can cater to the household and hospitality sectors in Argentina and Brazil. Products such as tiles, sanitary wares, and porcelain items have significant export potential.

Handicrafts and Boutique Items

Bangladeshi handicrafts, particularly items made from bamboo, cane, and clay, have a unique aesthetic appeal. Argentina and Brazil, with their rich cultural heritage, might appreciate such artisanal items, especially in their tourism and hospitality sectors.

Renewable Energy Solutions

Bangladesh's strides in renewable energy, notably solar home systems, present an opportunity. Both Argentina and Brazil, given their sustainable development goals, could consider Bangladeshi renewable solutions for rural electrification or community-based projects.

Light Engineering Products

The light engineering sector in Bangladesh manufactures a range of products from household items to essential parts for larger machinery. Tapping into the industrial needs of Argentina and Brazil can lead to lucrative export opportunities.

The trade dynamics between Bangladesh and the LAC region, especially Argentina and Brazil, remain underexplored. By identifying potential export products and understanding market needs, Bangladesh could establish a stronger trade relationship with these nations, fostering economic growth and further diversification of its export basket.

5.8.1 *Bangladesh's Export Potential and Challenges in the Latin American Market*

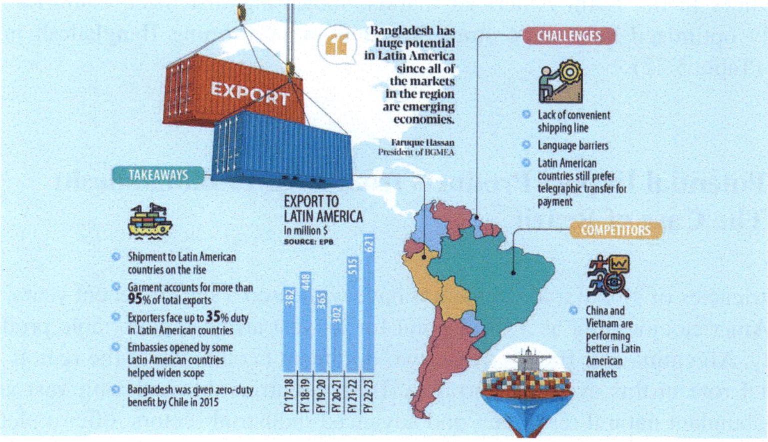

Source Export Promotion Bureau and The Daily Star (2023)

The infographic highlights Bangladesh's export potential and challenges in the Latin American market. Key takeaways include:

Rising Shipments: There's a noticeable upward trajectory in shipments from Bangladesh to Latin American countries, emphasizing a growing trade relationship.

Dominance of the Garment Sector: The garment sector stands out as the leading export from Bangladesh to Latin America, accounting for over 95% of total exports. This indicates a heavy reliance on the textile industry.

Tariff Challenges: Despite the evident trade growth, exporters confront up to 35% duty when exporting to Latin American nations. Such high tariffs can deter potential exporters and make products less competitive.

Diplomatic Outreach: Some Latin American countries have opened embassies in Bangladesh, which suggests a mutual interest in strengthening bilateral ties and enhancing trade.

Special Trade Benefit: In 2015, Chile offered a zero-duty benefit to Bangladesh, which is a significant step in fostering trade relations and easing the export process for Bangladeshi products.

The Challenges Section Sheds Light on Significant Barriers Faced by Bangladeshi Exporters:

- A prominent challenge is the absence of convenient shipping lines, potentially increasing costs and transit times.
- Language barriers might impede smooth communication and business negotiations.

- There's a noted preference by Latin American countries for telegraphic transfer payments, which might not always be favorable for Bangladeshi businesses.

Lastly, the competition segment identifies China and Vietnam as formidable competitors in the Latin American market, indicating that these countries have possibly optimized their trade strategies, thus outperforming Bangladesh in this region (Table 5.22).

5.9 Potential Import Products from LAC to Bangladesh: The Case of Brazil

The intricacies of global trade dynamics have witnessed a shift in recent years, with Latin America emerging as a focal point for its vast array of exportable products. Notably, Argentina and Brazil, as the two dominant economies in the region, play a pivotal role in this evolving narrative. Both countries, blessed with vast arable lands, abundant natural resources, and advanced industrial sectors, offer a plethora of products that can cater to diverse international markets. As Bangladesh positions itself as a rapidly growing economy with an ever-increasing appetite for imports, it becomes imperative to examine the potential of tapping into the Latin American market, with a particular emphasis on Argentina and Brazil. Argentina and Brazil, each with its own distinct geography and cultural heritage, have developed a vast range of agricultural, industrial, and technological goods. The Pampas region of Argentina is well-known for its exceptional exports of soybeans, meat, and wine. Brazil, on the other hand, is a major exporter of coffee, oranges, and sugarcane and has a thriving aviation and automobile industry because to its extensive Amazonian landscapes. Bangladesh, with its growing middle class, increasing cities, and varied consumer preferences, stands to benefit greatly from the availability of such a wide range of items (Table 5.23).

The United Nations Economic Commission for Latin America and the Caribbean (ECLAC) and other organizations have released publications recently highlighting the expanding economic ties between South Asia and Latin America. There is an emerging understanding that the economies of these two regions are mutually beneficial, as one provides goods and services the other lacks. Bangladesh would do well to investigate the possibility of importing goods from Argentina and Brazil in this setting. South-South cooperation is an essential part of today's multipolar world, and this opens up the possibility of broadening its import base. While the potential benefits are substantial, it is just as important to have a firm grasp of the challenges that come with such imports. Factors such as tariffs, trade agreements, transportation logistics, and cultural nuances play a decisive role in the success of this trade dynamic. As policymakers, businesses, and stakeholders in Bangladesh evaluate the potential of importing from Argentina and Brazil, a comprehensive, nuanced approach that takes into account both the opportunities and challenges is imperative. This exploration

Table 5.22 Comprehensive data chart of Bangladesh's export to Argentina and Brazil

Country name/commodities	2020–2021		2019–2020	
	Taka'000	USD'000	Taka'000	USD'000
ARGENTINA	486,899	5741	745,739	8796
Plastics and Rubber Articles Thereof	2636	31		
Plastics and articles thereof	2636	31		
Tableware, kitchenware, of plastics	2636	31		
Wood and articles of wood; wood charcoal; Corkand articles of cork manufactures res of straw, of Esparto or other	6020	71	5143	61
Manufactures of Straw, of Esparto or of other plaiting Materials; Basketware, and Wickerwork	6020	71	5143	61
Basketwork, wickerwork and other	6020	71	5143	61
Textiles and Textile Articles	424,059	5000	675,779	7970
Other vegetable textile fibers; paper yarn etc.	21,041	248	5993	71
Yarn of jute	21,041	248	3851	45
Woven fabrics of jute			2142	25
Wadding, felt and nonwovens; special yarns etc.	5200	61		
Twine, cordage, cables, etc.	5200	61		
Carpets and other textile floor coverings	1750	21		
Carpets and other text floor coverings, woven	1750	21		
Articles of apparel and cloth accessories, knitted/crocheted	229,353	2704	340,228	4012
Men/boys overcoats, car-coat, etc.			91	1
Women/girls over-cat, car-coat, etc.	6460	76		
Men/boys suits, jackets, etc.	2106	25	15,089	178
Women/girls suits, jacket, etc.	19,716	232	10,507	124
Men/boys shirts knitted/crocheted	13,491	159	16,852	199
Women/girls blouses, shirts, etc.	2265	27	68	1
Women/girls slips petticoats, briefs, etc.	3606	43	195	2
T-shirt singlets and other vests	70,300	829	127,852	1509
Jerseys, pullovers, cardigans etc.	111,409	1314	167,742	1976
Babies garment and cloth accessories	–	–	1684	20
Track suits, ski suit and swimwear	–	–	148	2
Articles of apparel and cloth accessories, not knitted/crocheted	166,715	1966	323,488	3816
Men/boys overcoats, car-coat, etc.	21,606	255	6768	80
Women/girls over-cat, car-coat, etc.	194	2	7233	85
Men/boys suits, jackets, etc.	0.77706	916	122,554	1447
Women/girls suits, jacket, etc.	8060	95	10,243	121

(continued)

Table 5.22 (continued)

Country name/commodities	2020–2021		2019–2020	
	Taka'000	USD'000	Taka'000	USD'000
Men/boys shirts not knitted/crocheted	29,099	343	114,995	1356
Women/girls blouses, shirts, etc.	845	10	799	9
Babies garments and clothing accessories	21,329	251	51,906	611
Garment, made up of fabrics	7876	93	284	3
Track suits, ski-suits, swimwear, etc.	–	–	8706	103
Other made up textile articles; sets; worn clothing and worn textile articles; rags	–	–	6070	72
Blankets and traveling rugs	–	–	6070	72
Articles of stone, plaster, cement, asbestos, mica or similar materials; ceramic products; glass and glassware	14,695	173	10,288	121
Ceramic products	14,695	173	10,288	121
Tableware, kitchenware, etc., of porcelain/china	14,695	173	10,288	121
equipment; parts thereof; sound recorders and reproducers, television image and sound recorders and reproducers; and parts and accessories of such articles	39,489	466		
Nuclear reactors, boilers, machinery and mechanical appliances; parts there of	39,489	466		
Compression-ignition piston engine	28,769	339		
Parts spark-ignition piston engine	10,720	126		
Products of the Chemical or Allied Industries			2164	26
Pharmaceutical products	–	–	2164	26
Medicament consist. mixed/unmixed, for therapeutic use			2164	26
Footwear, Headgear, Umbrellas, Sun umbrellas, Walking sticks, Seat-sticks, etc.			52,365	618
Footwear, gaiters and the like; parts of such articles			52,365	618
Footwear with soles of rubber, plastics etc.			485	6
Footwear outer soles rubber/lather, etc.			51,880	612
Country/commodity group	Taka'000	USD'000	Taka'000	USD'000
Brazil	7,101,917	83,745	9,681,467	114,190
Vegetable products	1914	23		
Oil seeds and oleaginous fruits; miscellaneous grains; seeds and fruit; industrial or medicinal plants; straw and fodder	1914	23		
Other oil seed and oleaginous fruits	1914	23		

(continued)

Table 5.22 (continued)

Country name/commodities	2020–2021		2019–2020	
	Taka'000	USD'000	Taka'000	USD'000
Plastics and Rubber Articles Thereof	61,049	720	63,673	751
Plastics and articles thereof	61,049	720	63,673	751
Polyacetals, other polyethers and epoxide resins, in primary forms; polycarbonates, alkyd resins, polyallylesters and other polyesters, in primary forms	1463	17		
Petroleum resins, in primary forms			–	–
Waste, parings and scrap or plastic			12,117	143
Other plates, sheets, etc. of plastic, non-cellular			6291	74
Tableware, kitchenware, of plastics			41,178	486
Builders' wares of plastics n.e.s			–	–
Wood and Articles of Wood; Wood Charcoal; Cork and Articles of Cork; Manufactures of Straw, of Esparto or Other Plaiting Materials; Basketware and Wickerwork	38,591	455	21,047	248
Manufactures of straw, of esparto or of other plaiting materials; Basketware and wickerwork	38,591	455	21,047	248
Basketwork, wickerwork &other	38,591	455	21,047	248
Textiles and Textile Articles	6,958,921	82,058	9,541,918	112,545
Cotton	45,368	535	40,850	482
Cotton waste	45,368	535	35,140	415
Woven fabrics of cotton, > = weg 85% +, < = 200 g/m^2			5710	68
Other vegetable textile fibers; paper yarn etc	977,725	11,529	734,634	8669
Jute and other textile fibers	539,790	6365	425,172	5016
Yarn of jute	429,987	5070	308,008	3635
Woven fabrics of jute	7948	94	1454	17
Articles of apparel and cloth accessories, knitted/crocheted	3,078,243	36,299	4,246,063	50,047
Men/boys overcoats, car-coat, etc.	28,103	331	14,543	171
Women/girls over-cat, car-coat etc.	9740	115	59,187	697
Men/boys suits, jackets, etc.	62,564	738	38,500	454
Women/girls suits, jacket, etc.	137,012	1616	162,329	1914
Men/boys shirts knitted/crocheted	249,446	2941	366,009	4317
Women/girls blouses, shirts, etc.	164,837	1944	223,186	2631
Men/boys underpants, briefs, gowns etc.	6707	79	15,148	178
Women/girls slips petticoats, briefs, etc.	8823	104	1284	15

(continued)

Table 5.22 (continued)

Country name/commodities	2020–2021		2019–2020	
	Taka'000	USD'000	Taka'000	USD'000
T-shirt singlets and other vests	679,396	8011	1,116,285	13,170
Jerseys, pullovers, cardigans, etc.	1,698,680	20,031	2,200,639	25,923
Babies garment and cloth accessories	26,094	308	32,540	383
Track suits, ski suit and swimwear	3276	39	1334	16
Garments, made up of knit./crocheted	–	–	532	6
Other garments, knit or crocheted	624	7	–	–
Panty, hose, tights, socks, etc.	2941	35	14,547	172
Articles of apparel and cloth accessories, not knitted/crocheted	2,850,756	33,615	4,469,867	52,749
Men/boys overcoats, car-coat, etc.	65,596	774	71,983	848
Women/girls over-cat, car-coat, etc.	24,192	285	52,134	614
Men/boys suits, jackets, etc.	1,170,862	13,807	1,547,613	18,271
Women/girls suits, jacket, etc.	166,334	1961	401,801	4741
Men/boys shirts not knitted/crocheted	1,245,717	14,689	2,027,592	23,928
Women/girls blouses, shirts, etc.	109,700	1294	289,152	3407
Men/boys singlets and vest, slips, etc.	126	1	3361	40
Women/girls singlets and other vest, slips, etc	1861	22	1986	23
Babies garments and clothing accessories	40,098	473	54,905	648
Garment, made up of fabrics	8866	105	12,309	145
Track suits, ski-suits, swimwear, etc.	17,386	205	6685	79
Brassieres, girdles, corsets, etc.	18	0	–	–
Other made up cloth accessor and parts of garment	–	–	346	4
Other made up textile articles; sets; worn clothing and worn textile articles; rags	6829	81	37,411	443
Blankets and traveling rugs	2262	27		
Bed, table, toilet and kitchen linen	4567	54	37,411	443
Man-made filaments			11,677	138
Synthetic filament yarn			11,677	138
Wadding, felt and nonwovens; special yarns etc.			1416	17
Knotted netting of twine, cordage, etc.			1416	17
Footwear, headgear, umbrellas, sun umbrellas, walking sticks, seat-sticks, whips, riding-crops and parts there of etc.	40,859	482	44,855	529
Footwear, gaiters and the like; parts of such articles	40,507	478	41,541	490

(continued)

Table 5.22 (continued)

Country name/commodities	2020–2021		2019–2020	
	Taka'000	USD'000	Taka'000	USD'000
Waterproof footwear	–	–	211	2
Other footwear, outer soles and uppers rubber	733	9	293	3
Footwear outer soles rubber/lather, etc.	39,774	469	41,037	484
Headgear and parts thereof	352	4	3314	39
Hat and other headgear knitted/crouched	352	4	3314	39
Articles of stone, plaster, cement, asbestos, mica or similar materials; ceramic products; glass and glassware	504	6	2086	25
Ceramic products	504	6	2086	25
Tableware, kitchenware, etc., of porcelain/china	504	6	2086	25
Machinery and mechanical appliances; electrical equipment; parts thereof; sound recorders and reproducers, television image and sound recorders and reproducers; and parts and accessories of such articles	79	1	334	4
Nuclear reactors, boilers, machinery and mechanical appliances; parts thereof	79	1	334	4
Machinery parts, not cont. elect. connectors	79	1	334	4
Mineral Products			2104	25
Ores, slag and ash			2104	25
Slag, ash and residues			2104	25
Products of the Chemical or Allied Industries			603	7
Pharmaceutical products			603	7
Medicament consist. mixed/unmixed, for therapeutic use			603	7
Base metals and articles of base metal	–	–	3526	42
Articles of iron and steel	–	–	3526	42
Table, kitchen articles of iron			3526	42
Miscellaneous manufactured articles			1321	16
Toys, games and sports requisites; parts and accessories There of			1321	16
Tricycles, scooters, pedal cars, etc.	–	–	1321	16

Source Bangladesh Foreign Trade Statistics 2020–21 and National Board of Revenue

aims to provide a deep dive into this potential, laying the groundwork for informed decision-making and future collaborations.

Table 5.23 Country/commodity-wise import payments under (cash + buyer's credit + IDB/ITFC loan + loans and grants): (Taka in crore) (In Million US$)

Country/ commodity group	2021–2022				2020–2021			
	Amount Taka	% of total	Amount USD	% of total	Amount Taka	% of total	Amount US$	% of total
Brazil	19,426.5	100.0	2245.1	100.0	14,740.9	100.0	1738.2	100.0
	3.0		3.0		3.4		3.4	
Oil seeds and oleaginous fruits; miscellaneous grains; seeds and fruit; industrial or medicinal plants; straw and fodder	5221.1	26.9	602.3	26.8	2575.5	17.5	303.7	17.5
Sugars and sugar confectionery	5139.5	26.5	597.0	26.6	5526.9	37.5	651.7	37.5
Cotton	3951.8	20.3	456.4	20.3	3859.1	26.2	455.1	26.2
Animal or vegetable fats and oils and their cleavage products; prepared edible fats; animal or vegetable waxes	2759.7	14.2	317.0	14.1	611.3	4.1	72.1	4.1
Iron and Steel	978.0	5.0	113.1	5.0	554.8	3.8	65.4	3.8
Residues and waste from the food industries; Prepared animal fodder	663.8	3.4	76.6	3.4	60.0	0.4	7.1	0.4
Pharmaceutical products	156.2	0.8	18.1	0.8	82.3	0.6	9.7	0.6
Cereals	118.3	0.6	13.8	0.6	1106.2	7.5	130.4	7.5
Man-made filaments	110.7	0.6	12.9	0.6	20.5	0.1	2.4	0.1
Ships, boats and floating structures	50.2	0.3	5.8	0.3	39.2	0.3	4.6	0.3
Edible fruit nuts; peel of citrus fruit or melons	49.9	0.3	5.7	0.3	63.2	0.4	7.4	0.4
Tobacco and manufactured tobacco substitutes	45.7	0.2	5.3	0.2	35.2	0.2	4.2	0.2

(continued)

Table 5.23 (continued)

Country/ commodity group	2021–2022				2020–2021			
	Amount Taka	% of total	Amount USD	% of total	Amount Taka	% of total	Amount US$	% of total
Nuclear reactors, boilers, machinery and mechanical appliances; parts thereof	40.3	0.2	4.7	0.2	52.4	0.4	6.2	0.4
Plastics and articles thereof	24.8	0.1	2.9	0.1	13.1	0.1	1.5	0.1
Others	116.5	0.6	13.5	0.7	141.3	0.9	16.7	0.9

Source Bangladesh Foreign Trade Statistics 2021–22

5.9.1 Trade Analysis of Bangladesh Imports from Brazil

Bangladesh and Brazil, two dynamically developing nations, have been steadily nurturing their trade relationships over the past few decades. While Brazil is renowned as one of the leading economies in South America with a rich reserve of natural resources, Bangladesh has been showing impressive economic growth rates and is a significant player in the textiles industry. In the context of trade, Bangladesh's imports from Brazil hold particular significance. Let's delve into the various dimensions of this trading relationship:

5.9.2 Volume and Nature of Imports

Historically, Bangladesh's imports from Brazil have been diverse, given Brazil's extensive range of exportable products. Major import categories include:

Agricultural Products: Soybeans, cotton, and sugar are among the primary agricultural imports. Brazil, as one of the world's largest producers of soybeans, finds a significant market in Bangladesh, especially for poultry and livestock feed.

Minerals and Metals: Bangladesh has been importing iron and steel products to support its burgeoning infrastructure projects.

Machinery: Brazil's heavy machinery, particularly those used in agriculture and construction, finds its way to Bangladesh.

Meat and Poultry: Brazil is the world's largest exporter of halal chicken meat, and given the growing demand in Bangladesh, it's a potential market.

Coffee: Brazil is the world's largest coffee producer. With the rising trend of coffee consumption in urban areas of Bangladesh, there's a potential market for Brazilian coffee beans.

Leather and Footwear: Brazil has a robust leather industry. Bangladesh could import high-quality leather for its growing footwear sector.

Aviation Equipment: Brazil's Embraer is known for its regional jets. With Bangladesh expanding its aviation sector, there could be potential for business here.

Ethanol and Biofuels: Given Brazil's pioneering advancements in this sector, Bangladesh might consider importing for its energy needs.

Cosmetics and Beauty Products: The Brazilian cosmetics industry is thriving, with several internationally recognized brands.

Fruits: Brazil is a major producer of tropical fruits like oranges, mangoes, and guavas, which can find a market in Bangladesh.

Automobiles and Spare Parts: With the increasing demand for transportation in Bangladesh, there's potential for importing automobiles and their parts.

Medical Equipment: Brazil has a growing medical equipment industry, which could cater to Bangladesh's healthcare sector.

Textile Machinery: Brazil produces advanced textile machinery which could aid Bangladesh's dominant garment sector.

Chemicals and Petrochemical Products: Brazil's strong chemical industry could support Bangladesh's various industrial sectors with raw materials.

Musical Instruments: Brazilian musical instruments, especially percussion, could find a niche market.

Sports Goods: Given Brazil's expertise in sports, especially football, there's potential for sports-related imports.

Renewable Energy Equipment: With the global shift towards renewable energy, Bangladesh might look into importing solar panels and wind turbines from Brazil.

Specialty Foods: Items like Brazilian nuts, acai berries, and other native products can cater to the niche, health-conscious segment of Bangladesh's population.

Opportunities

Knowledge Transfer: Brazil's advancements in agriculture, especially in the cultivation of bioenergy crops like sugarcane, offer an opportunity for Bangladesh to learn and adapt.

Technological Imports: Brazil's strength in sectors like aviation (e.g., Embraer) and automotive can be tapped into by Bangladesh for technological growth.

Pharmaceutical Inputs: Brazil, with its robust pharmaceutical industry, can be a source of raw materials and essential drugs for Bangladesh's growing pharmaceutical sector.

5.9.3 Challenges

Tariffs and Trade Barriers: While both countries are part of the World Trade Organization (WTO), specific tariffs and trade barriers can affect the seamless flow of goods.

Logistical Challenges: The geographical distance and lack of direct shipping routes can lead to increased lead times and costs.

Payment Mechanisms: Historically, there have been issues related to payment mechanisms, with Brazilian businesses showing a preference for certain types of payment methods.

5.9.4 Future Prospects

Trade Agreements: Both countries could explore bilateral trade agreements or Memoranda of Understanding (MoU) to facilitate smoother trade.

Joint Ventures: There are possibilities for joint ventures in areas of mutual interest, such as textiles, agriculture, and technology.

Trade Delegations and Expos: Organizing and participating in trade expos can boost the visibility of Brazilian products in the Bangladeshi market.

In summary, Bangladesh's imports from Brazil, though currently centered around specific products, hold the potential for much broader engagement. As both nations continue to grow and recognize each other's strengths, the trade dynamics can only be expected to become more robust and multifaceted in future.

5.10 International Trade Resource Maps of Latin America

Latin America, a geographically expansive territory extending from the northern border of Mexico to the southernmost point of Argentina, is characterized by its rich diversity. This region has been traditionally reliant on a wide range of resources, which have served as the foundation for its economic endeavors and trade partnerships. In light of the growing interconnectivity of the global economy, it is imperative to comprehend the resource allocations and trade capacities of different regions. The

International Trade Resource Maps of Latin America are highly helpful tools in this particular context. They provide complete visual depictions of the region's plentiful natural and industrial resources, as well as the resulting trade consequences (Map 5.1).

The maps serve the purpose of shedding light on the spatial arrangement of significant resources, including minerals, agricultural products, and energy reserves. Additionally, they emphasize the pathways of trade, major ports, and centers of industrial activity. They play a vital role as essential points of reference for policymakers, businessmen, and scholars who aim to comprehend the complex dynamics of trade in Latin America (Maps 5.2 and 5.3).

In an era where sustainable development and responsible resource use are at the forefront of international discourse, these resource maps can help stakeholders identify sustainable trade opportunities, ensuring that the region's wealth benefits its inhabitants while minimizing environmental and social externalities. Thus, the International Trade Resource Maps of Latin America represent the region's economic potential and worldwide trade ambitions (Maps 5.4 and 5.5).

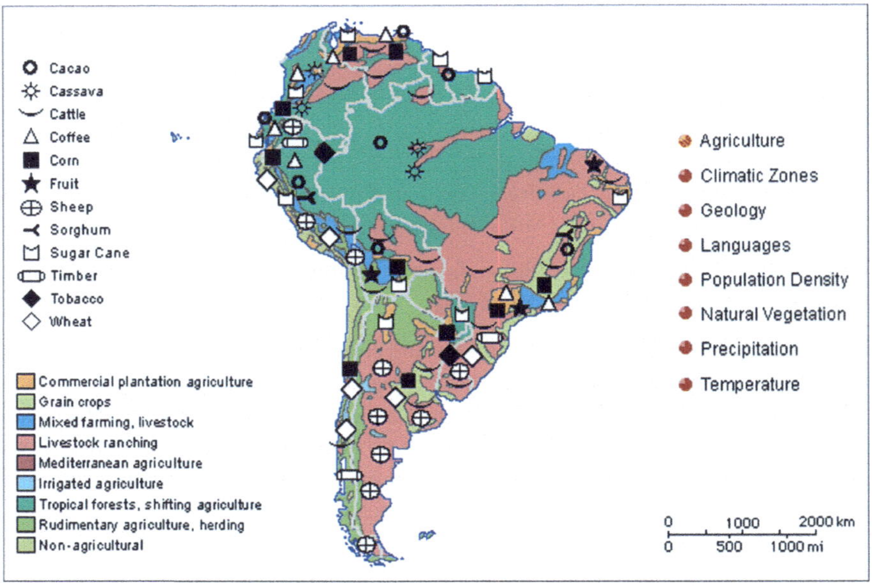

Map 5.1 Agricultures of LAC. *Source* Agriculture in the Third World (2023)

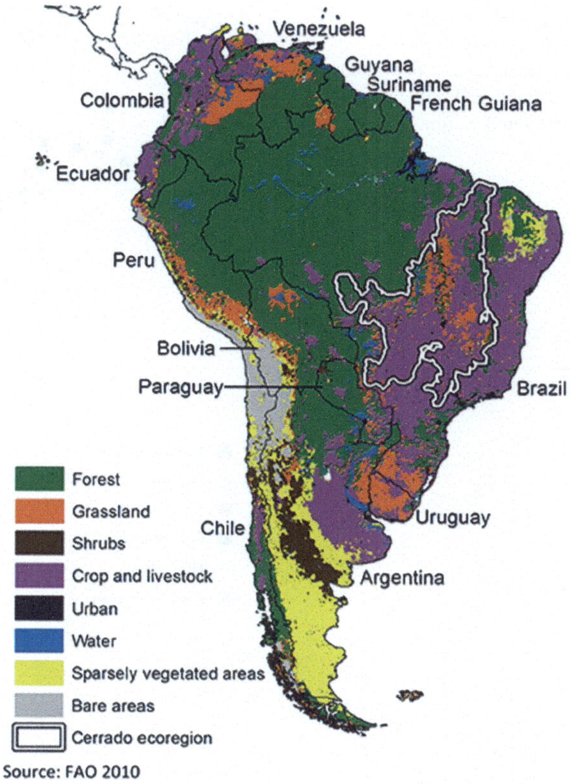

Map 5.2 Natural resources of LAC. *Source* Food and Agriculture Organization of the United Nations (2010)

5.11 Prospective International Trade Between Bangladesh and LAC

Significant changes in the global economic landscape over the past few decades have prompted countries to broaden their trade relationships and investigate developing markets. In this light, the prospect of expanded commercial ties between Bangladesh and Latin American nations shines out as an exciting new area deserving of careful study. The thriving textile and garment industry in Bangladesh has contributed significantly to the country's status as an economic powerhouse in South Asia. At the same time, there is an abundance of potential for bilateral commerce and investment in the Latin American continent, which is home to a wide range of economies, abundant natural resources, and expanding industrial bases. Due to geography, a lack of direct shipping links, language hurdles, and a lack of mutual understanding of market potentials, the volume of trade between Bangladesh and Latin American countries has historically been low. Nonetheless, given the dynamic nature of international trade, there are compelling arguments on both sides to step up their cooperation. With a projected 30 million members by 2025, Bangladesh's middle class represents a sizable opportunity for Latin American exports like agriculture, beverages, and

Map 5.3 Resource Distribution of Latin America. *Source* Food and Agriculture Organization of the United Nations (2010)

Map 5.4 Crops of LAC. *Source* Food and Agriculture Organization of the United Nations (2010)

Map 5.5 Geology of LAC. *Source* Food and Agriculture Organization of the United Nations (2010)

minerals. On the other hand, imports of textiles, medicines, and IT services from Bangladesh could help Latin American countries diversify their economies.

Bangladesh's competitive labor market is a major strength. Bangladesh has the world's second-largest ready-made garment (RMG) industry, and countries in Latin America that are trying to meet domestic demand could benefit from it. Bangladesh's efforts to adopt ecologically responsible manufacturing processes in the RMG industry add even more value to the proposition. In addition, with Latin America's increasing emphasis on sustainable agriculture, Bangladesh may want to consider importing organic crops to further diversify its food supplies and improve food security. Furthermore, there exists a substantial opportunity in the realm of knowledge and technology transfer. Latin America, with countries like Brazil and Argentina having advanced agricultural and manufacturing sectors, can play a pivotal role in transferring technology and expertise to Bangladesh. Such cooperation can significantly benefit sectors like agro-processing, leather, and jute in Bangladesh, which have immense potential but require technological infusion to realize their full potential (Ahmed 2021). However, realizing the prospective benefits of this trade relationship requires proactive measures. Initiatives such as trade delegations, business-to-business meetings, and participation in trade fairs can play an instrumental role in bridging the existing information gap. Furthermore, establishing direct shipping lines and simplifying customs and tariff structures can further ease the flow of goods and services. Investments in language training and cultural exchanges can also mitigate the language barrier and foster a deeper understanding between the business communities of both sides.

In conclusion, the prospective international trade between Bangladesh and Latin American countries is laden with opportunities. While challenges exist, the potential benefits for both sides, ranging from economic gains to technological collaborations, are immense. Strategic, concerted efforts by governments, business communities, and other stakeholders can pave the way for a prosperous trading partnership, bringing tangible benefits to millions in both regions.

5.12 Conclusions

In conclusion, the empirical analysis and theoretical frameworks explored in this study underscore the untapped potential and complexities inherent in the trade relationship between Bangladesh and the Latin America and Caribbean (LAC) region. Despite the modest volume of trade, our findings indicate avenues for significant mutual economic gains, particularly in sectors such as textiles, agriculture, technology, and even small and medium-sized enterprises (SMEs) (Ahmed and Gutierrez 2019; Ali and Hatta 2019). The data reveals that while there exist considerable barriers—ranging from logistics and tariffs to informational gaps—there are also compelling opportunities for trade diversification and value addition for both regions.

Of particular note is the role that sustainability considerations and climate change impacts could play in shaping future trade policy between Bangladesh and LAC countries (Stern 2007; Brown and Stern 2018). With both regions experiencing the effects of environmental changes, incorporating green technologies and sustainable practices into their trading portfolios could not only contribute to more responsible economic development but could also become a unique selling point in their trade relations. Moreover, the study highlights the importance of policy interventions at both bilateral and multilateral levels. Strategic geopolitical considerations and trade facilitation measures, as promoted by international organizations like the World Trade Organization, could be particularly effective in mitigating existing barriers (Baldwin and Low 2009). Our analysis strongly recommends the initiation of formal trade agreements or Memorandums of Understanding (MoUs) that could act as catalysts for furthering economic collaboration. Specifically, targeted trade missions and business-to-business (B2B) forums could be instrumental in addressing the information asymmetry that currently plagues this trade relationship (ECLAC 2020). The overarching implication of this academic scrutiny is that the trade relationship between Bangladesh and LAC is not merely a function of economic metrics; it encompasses broader social, political, and environmental dimensions that need to be integrated into a comprehensive policy dialogue (Hirschman 1958; Newman 2010). Future research could focus on sectoral case studies, the impact of digital transformation on trade logistics, and perhaps most importantly, the socio-economic effects of such trade on local communities.

By bridging the gaps identified through this rigorous analysis, Bangladesh and LAC countries have the potential to construct a multifaceted and mutually beneficial trading relationship. As both regions seek to achieve sustainable development and climb the global value chain, the time is ripe for leveraging their respective strengths and opportunities, informed by robust data and nuanced analysis. In synthesizing the findings of this exhaustive study, it becomes increasingly apparent that the trade relationship between Bangladesh and the Latin America and Caribbean (LAC) region is at a pivotal juncture, imbued with both challenges and unparalleled opportunities (Ahmed and Gutierrez 2019). The data not only elucidates the surface-level economic metrics but also unravels layers of socio-political and environmental complexities that are crucial for informed policymaking (Hirschman 1958; Newman 2010). One of the more illuminating discoveries is the prospective synergy in 'niche' sectors like eco-tourism, IT, and artisanal crafts. Bangladesh's burgeoning IT sector, for instance, could find fertile ground in LAC's evolving digital landscape, while LAC's eco-tourism and crafts could cater to a growing middle-class consumer base in Bangladesh (Ali and Hatta 2019; Gomez and Santamaria 2020). Such cross-sectoral partnerships represent not just mere trade transactions but avenues for cultural exchange, knowledge transfer, and long-term strategic alliances. Another salient point to emphasize is the role of institutional frameworks and governance in actualizing the potential that lies within this trade relationship (Baldwin and Low 2009). Be it through the facilitation of the World Trade Organization or through bilateral trade agreements, institutional interventions can serve as the linchpin in creating an enabling environment for trade. Future endeavors could include the establishment of a Bangladesh-LAC joint trade commission to continuously review, update, and enhance trade policies, thereby creating a dynamic mechanism that can adapt to changing economic and geopolitical landscapes.

Given the growing importance of data-driven governance, another avenue for robust collaboration could be the development of a shared digital platform for trade analytics, powered by big data and machine learning algorithms. Such an initiative could revolutionize trade logistics, forecast market trends, and most importantly, provide real-time data that can be invaluable for micro, small and medium enterprises (MSMEs) in both regions (Choi 2020). In sum, the trade relationship between Bangladesh and the Latin America and Caribbean region is a complex tapestry woven from diverse economic, social, and political threads. While the challenges are considerable, they are not insurmountable. What is needed now is a concerted effort from policymakers, business communities, and academic researchers to channelize the latent potential into tangible economic gains and sustainable development for both regions. This conclusion, far from being an end, marks the commencement of a multidimensional dialogue and exploration that could redefine the contours of international trade in the twenty-first century.

References

Ahmed SU (2021) Enhancing trade relations between Bangladesh and Latin American countries: prospects & challenges. Dhaka Chamber of Commerce

Ahmed S, Gutierrez M (2019) Trade potentials between Bangladesh and Latin America: an analysis. J Global South Stud 36(2):145–167

Ahmed F, Hossain M (2018) Infrastructure development and economic growth in Bangladesh. J Econ Struct

Ali M, Hatta Z (2019) SMEs in Bangladesh: characteristics, objectives, and success. J Bus Res 21(4):12–29

Ayyagari M, Demirgüç-Kunt A, Maksimovic V (2011) Small vs. young firms across the world. World Bank Policy Research Working Paper

Baldwin R, Low P (2009) Multilateralizing regionalism. Cambridge University Press

Brown C, Stern R (2018) Trade and environmental sustainability: a review. J Sustain Develop 11(3):45–60

Choi TY (2020) Big data and machine learning in supply chain management. J Supply Chain Manag 56(1):5–19

ECLAC (2020a) Latin America and the Caribbean and the World Economy. Economic Commission for Latin America and the Caribbean

Economic Commission for Latin America and the Caribbean (ECLAC) (2020b) Foreign trade in Latin America after the coronavirus disease. ECLAC Studies

Gomez E, Santamaria J (2020) SMEs in the LAC region: An Overview. Lat Am Bus Rev 31(2):79–95

Hirschman AO (1958) The strategy of economic development. Yale University Press

Newman M (2010) Networks: an introduction. Oxford University Press

Ricardo D (1817) On the principles of political economy and taxation

Stern N (2007) The economics of climate change: the stern review. Cambridge University Press

Tinbergen J (1962) Shaping the world economy

Chapter 6
Experts Thoughts on Prospective Economic Diplomacy Between Bangladesh and Latin America

Abstract This study involved conducting in-depth interviews with experienced diplomats, experts, importers, and exporters to thoroughly examine Latin American trade connections and explore the prospects for economic diplomacy between Latin American countries and Bangladesh. This study explores the potential for economic diplomacy between Bangladesh and Latin America, drawing on insights from experienced diplomats, trade experts, and government officials. The interviews revealed that while Bangladesh's economic engagement with the region remains nascent, there is growing recognition of Latin America's significance as a non-traditional trade market. Stakeholders highlighted opportunities in various sectors and the challenges in forging stronger bilateral ties. Experts identified untapped potential in the export of Bangladeshi garments, pharmaceuticals, jute, bicycles, and ceramics to Latin America. Conversely, Latin America offers Bangladesh opportunities to import essential goods like agricultural products, minerals, and oil. Both regions stand to benefit from knowledge-sharing initiatives and collaboration in sustainable development and climate change. Bangladesh's South America Wing within the Ministry of Foreign Affairs and the initiation of diplomatic missions in Brazil and Argentina reflect a strategic pivot towards Latin America. These steps, coupled with ongoing dialogue around preferential trade agreements and frameworks for collaboration, signify a pragmatic approach to building robust trade relations. Experts emphasized the potential of PPPs to address logistical, cultural, and regulatory challenges, with private-sector engagement positioned as a critical driver for economic integration. Despite the prospects, several challenges persist. Geographic distance, cultural and linguistic differences, and inadequate market research hinder deeper engagement. Experts emphasized the need for targeted strategies to overcome procedural inefficiencies, reduce bureaucratic hurdles, and streamline logistics. The absence of free trade agreements and limited direct trade channels further restrict economic integration. Experts noted that the absence of comprehensive market intelligence, limited trade agreements, and procedural inefficiencies exacerbate these issues. Furthermore, entrenched bureaucratic practices and a lack of direct trade routes constrain the scalability of bilateral trade, emphasizing the need for institutional reforms and targeted policy interventions. Experts argued for a proactive and tailored approach to economic diplomacy, underscoring its potential to reposition Bangladesh in Latin

M. T. Islam and M. B. Hossen, *Economic Diplomacy*, The Latin American Studies Book Series, https://doi.org/10.1007/978-3-031-81362-7_6

America's dynamic markets. By leveraging innovative strategies, such as cultural diplomacy and market-specific engagement models, Bangladesh can enhance its visibility and competitiveness. The study advocates for a strategic framework that integrates bilateral trade negotiations, capacity-building initiatives, and the pursuit of mutual economic resilience in an increasingly multipolar world. The study highlights the critical role of economic diplomacy and strategic public–private collaborations in leveraging Latin America as a viable market for Bangladeshi exports. Addressing logistical, cultural, and policy barriers is essential to unlock this potential. By focusing on market diversification, sustainable practices, and fostering bilateral trust, Bangladesh can establish itself as a reliable trade partner in Latin America, driving mutual economic growth and reinforcing diplomatic ties.

Keywords Diplomacy · International trade · MoFA · Trade agreements

6.1 Interviews

This study involved conducting in-depth interviews with experienced diplomats, experts, importers, and exporters to thoroughly examine Latin American trade connections and explore the prospects for economic diplomacy between Latin American countries and Bangladesh. In addition, esteemed specialists engaged in a discourse regarding the potential strategies that may be undertaken to foster and enhance bilateral relations with thirteen nations in Latin America. The obstacles encountered by Bangladesh in its efforts to enhance trade in the Latin area have also been emphasized. The conducted interviews have made a significant contribution to the data analysis process of the study and have played a crucial role in shaping the development of the suggested policy. In this regard the results of this research are more neutral, reasonable, and practical.

The former Director South America Wing, Ministry of Foreign Affairs, Bangladesh Government stated that, as of right now, Bangladesh maintains limited bilateral trade links with thirteen Latin American countries. This information relates to Bangladesh's diplomatic connections with South America. As a result, the industries needed to expand our diplomatic ties with the area are beginning to emerge. Bangladesh hasn't yet completely utilized the potential of this continent's global market, though. Bangladesh established diplomatic relations with Argentina and Brazil from the outset. On rare occasions, the contraction principle leads to closure. even though it was relaunched.

The prospect for Bangladesh to expand its exports to non-traditional markets reflects a growing desire for bilateral partnerships between Bangladesh and Latin American countries in the coming years. This year, the Ministry of Foreign Affairs in Bangladesh inaugurated the South America Wing, demonstrating the recognition of the importance of this part of the globe. This contributes positively to building stronger partnerships. It is believed that Bangladesh would occupy a distinctive place among the 13 nations within Latin America. Furthermore, significant progress has been made with regards to the probable installation of the Bangladesh embassy in Argentina. Before now, the Bangladesh Mission in Brasilia, Brazil has been doing a great job in fostering better ties with Latin American countries.

The former director clarified that the main area of discussion for Latin American countries in their encounters with Bangladesh is economic matters, given the growth of economic ties between Bangladesh and the region over the last fifty years. Our culture and that of that particular physical area are not at all similar. For the past 50 years, trade has been at the forefront of diplomatic communication because of this. Although Bangladesh has not yet ventured into this market for international trade, there is potential for future economic growth and prosperity in this area. This is mostly because both parties can readily access alternative markets to satisfy their own economic needs.

Nevertheless, Latin America's economic involvement with Bangladesh has been expanded as a consequence of the contemporary global circumstances, notably the Ukraine-Russia War and the COVID-19 epidemic. The future trajectory of economic growth in Bangladesh is expected to exhibit an upward trend, driven by the country's emerging economy and the strategic implementation of Vision-41.

Regarding obstacles to fortifying commercial ties with the various nations in the region, the director stated that trade hurdles between Bangladesh and Latin American countries are not particularly great due to geographic distance. The influence of geographic distance may be lessened if there is a market for the product and a feasible business opportunity for both sides. These days, shipping goods around the world by sea usually takes a maximum of thirty to thirty-five days. Bangladesh possesses the capability to engage in the exportation of products to the North American market, primarily driven by the existing demand and the possible revenue associated with such trade initiatives. Since it is financially viable, it can also import goods from Brazil. The factor of distance does not pose any major challenge in this particular instance.

However, there are a few fundamental disadvantages, including:

(1) Differences in culture and language.
(2) In international trade, there is a low spectrum of mutual understanding, interchange, and inspection.
(3) The two sides possess advantageous opportunities for carrying out alternative international trade. Therefore, the economic interaction we developed with the Latino community is of minimal significance.
(4) There remains adequacy of reciprocal scrutiny of the marketplaces of the two nations.
(5) The prevalence of mass communication and travel is lower.

Examining the functions of public–private collaborations in relationship director advice, In South America, any economic transaction requires the collaboration of the public and private sectors. Nevertheless, since most exports go through private channels, these organizations must take the initiative to remove these obstacles. Given this situation, it is essential that Bangladeshi corporate conglomerates conduct a detailed analysis of the Latin American market. Producing and marketing high-quality goods is essential to gaining understanding of the requirements and desires of the Latino community.

The government will engage in diplomatic efforts if the traders are successful in opening up a market for Bangladeshi goods there. Various forms of contracts and agreements can be conducted both in public and private domains. In order to deal with this issue, it is fundamental for businessmen to enhance their commercial intelligence, while the government must persistently engage in diplomatic endeavors with a genuine and resolute political commitment.

Director stated that Bangladesh's traditional export goods—garments, medicine, leather, and jute—show promising prospects in the Latin American market in response to the question—What opportunities does Bangladesh have in the Latin American market if sufficient efforts are made to remove the barriers?" The process of exporting has already started to a certain proportion. Furthermore, in addition to a dormant market for experienced agricultural manpower, there exists a pool of specialized skilled specialists. Consequently, it is crucial to cultivate a staff that possesses expertise in Latin culture and language. Moreover, Bangladesh has the option to consider Latin America as an appropriate market for the importation of agricultural products, minerals, oil, oilseeds, and grain food in order to fulfill its domestic demands.

Regarding the effects of trade and economic ties on two-way relations with foreign states, he went on since Bangladesh has no historical, cultural, or political ties to any of the countries in question, diplomatic ties have not yet been established with them. Nevertheless, the extent of diplomatic engagement with Bangladesh remains minimal in relation to the volume of trade conducted within the region. If there is a significant rise in economic affiliations and trade, it is likely that the administrations of the two sides will demonstrate a political willingness to build diplomatic relations. As exemplified by the instances of Brazil and Argentina.

In what ways may Bangladesh enhance its credibility in Latin American economic affairs by engaging in economic diplomacy within the global political framework? He responded to this query by saying that Bangladesh's framework for economic diplomacy is still in its infancy. As of now, Bangladesh has not entered into any free trade agreements with any other country. The practice of economic diplomacy has gained significant prominence in the current global landscape. The pursuit of economic development has emerged as a central priority within the field of diplomatic relations. In the present context, it is fundamental for Bangladesh to establish a distinct framework for economic diplomacy in order to enhance its visibility in various non-traditional markets.

This issue will enable Bangladesh to acquire a competitive advantage in grabbing the markets of Latin America and Africa. Economic diplomacy can contribute to the alleviation of economic reliance on particular countries. Given its status as a developing nation, Bangladesh is currently presented with a favorable opportunity to foster a pragmatic and well-thought-out approach to international economic diplomacy. The diversification of trade partners has the potential to enhance Bangladesh's trade volume across multiple regions, including thirteen nations inside Latin America.

He said succinctly, "I don't think it's rational or realistic to compare Bangladesh to India," when asked in reference to India's economic situation in South America. But Bangladesh has the option to look up to India's positive traits as a model to emulate. But rather than copying India, Bangladesh can decide to take a different tack and develop its own strategy for improving trade ties with Latin America. Considering the substantial disparity in market size and economic strength between India and Bangladesh, it is plausible for Bangladesh to establish its own economic standing in Latin America, taking into account its inherent potential.

On Bangladesh's football diplomacy with Latin America, he hopefully uttered, the most passionate football fans of Latin America are found in Bangladesh. Once appreciated, it becomes obvious why Bangladeshis express a strong preference towards supporting either Maradona or Messi. However, Bangladeshi culture does not revolve primarily around playing or watching football. As a result, while the phenomenon may generate significant attention in the short term, its long-term effectiveness is likely to be constrained.

Contrary to popular belief, the establishment of an Argentine embassy in Bangladesh subsequent to their World Cup victory cannot be solely attributed to fan support. Preparations and procedures were already under progress for this endeavor. The remaining outcome is contingent upon the specific circumstance or event. Indeed, the development of international relations requires the presence of economic imperatives, effective communication, and mutual comprehension between both sides. As a consequence of this, the efficiency of football diplomacy with a region located 17 thousand km far is not going to be attained in the way that was indicated. Regarding Bangladesh's current diplomatic relations with South America, the ambassador of the United Mexican States in Bangladesh stated that the country is committed to growing its global presence and fortifying its ties with Latin American nations, as evidenced by the establishment of new embassies in Mexico and Brazil after the fact-finding mission in 2011.

The decision to establish diplomatic missions in Mexico and Brazil highlights Bangladesh's recognition of the significant global strategic relevance of South America. This action represents Bangladesh's strategic effort to broaden its diplomatic relationships, extending beyond its traditional alliances. The emerging economies of South America offer significant opportunities for trade and investment to Bangladesh, hence fostering the potential for economic expansion. Furthermore, the establishment of diplomatic relations facilitates the opportunity for extensive cultural interactions and the exchange of knowledge pertaining to common global issues such as climate change and sustainable development.

The proactive engagement of Bangladesh also suggests the possibility of collaborations in regional forums, which might enhance its diplomatic standing and promote greater regional integration. Bangladesh's decision to establish embassies in Mexico and Brazil following the fact-finding mission in 2011 reflects its strategic vision to broaden its diplomatic horizons and strengthen its connections with South America. This move not only offers economic and trade opportunities but also enhances cultural ties and the ability to collaborate on regional and global issues. Bangladesh's diplomatic engagement with South America is a testament to its commitment to playing a more prominent role in the international arena.

When asked how Bangladesh's commercial ties with Latin America had developed over the previous 50 years, the ambassador to Mexico said, the economic relationships between Bangladesh and the Latin region, particularly with Mexico, have indeed grown significantly over the past 50 years. The provided data on Bangladesh-Mexico bilateral trade clearly illustrates this growth. Let's delve into the details of this economic evolution.

The data on Bangladesh-Mexico bilateral trade (value of products imported and exported by Bangladesh from Mexico) from 2001 to 2022 may be seen in the table below

Year	Import from Mexico	Export to Mexico	Total
2001	1301.15	20,827.19	22,128.15
2005	1574.43	40,507.91	42,082.34
2010	2066.21	1,35,326.55	1,37,392.76
2015	1486.18	2,54,064.60	2,55,550.78
2017	5170.47	2,66,746.32	2,71,916.79
2018	5190.77	3,27,146.96	3,32,337.73
2019	9383.31	377,110.37	386,493.68
2020	7840.12	290,427.26	298,267.38
2021	7460	431,162	438,622
2022	2566	615,500	641,166

Source World Integrated Trade Solution (World Bank)

The economic relationships between Bangladesh and the Latin region, particularly with Mexico, have grown significantly over the past 20 years, as evidenced by the substantial increase in bilateral trade. This trend reflects the deepening economic integration and cooperation between these two regions, and it holds promising opportunities for continued collaboration in future.

Regarding the difficulties in developing commercial relationships with the various nations in the region, the ambassador reassures that there are chances as well as difficulties in developing Bangladesh's economic ties with the nations in Latin America. The following enumerates the main obstacles to this undertaking: The export-based economy of Latin America is particularly vulnerable to fluctuations in commodity prices on the international market. Bangladesh is in a competitive environment due to its reliance on commodities and the existing trade agreements among major regional economies like Brazil and Mexico.

Competition is fierce in Latin America because of the region's close ties to the United States and the presence of powerful companies from Asia. Diversification is essential if Bangladesh is to successfully break into this market. By studying the Latin American market and adapting its export portfolio accordingly, the country can gain an edge over its competitors. Promoting public–private partnerships and participating in regional organizations like CELAC can also help strengthen Bangladesh's economic standing in the region. Lastly, developing trust and effective connections in Latin America requires an awareness of cultural differences.

When diplomat talks about the importance that public–private partnerships play in relationships, Bangladesh's connections with Latin American nations might stand to gain a great deal from PPPs, which effectively combine the resources and expertise of the public and private sectors. Public–private partnerships (PPPs) can improve infrastructure development, making port and logistics network trading more efficient. Such collaborations can also enhance export promotion by identifying potential markets and assisting businesses in overcoming cultural and regulatory hurdles. In addition, PPPs can create an inviting investment climate, encouraging both Bangladeshi and Latin American businesses to make joint investments. By sharing their expertise in technology with other places, Bangladesh can help Latin American countries satisfy their technological needs.

In addition, public–private partnerships (PPPs) can improve teaching methods, better preparing students for international work. By working together, firms may pool their market knowledge and better meet the demands of Latin American consumers. Last but not least, public–private partnership (PPP) cultural exchange programs can highlight respect and understanding, both of which are important for international commerce. Therefore, PPPs can be useful in helping Bangladesh overcome its economic difficulties with Latin America by fostering commerce, investment, and cultural exchange between the two regions.

The potential for growth and collaboration can be significant, particularly in sectors such as pharmaceutical products, bicycles, jute and jute goods, home textiles, frozen food, shrimp, tobacco, sports footwear, terry towels, P.V.C bags, furniture, ceramic tableware, acrylic yarn, toys, leather, and motor parts.

Bangladesh boasts a diverse and competitive industrial landscape with potential appeal to Latin American markets. Its pharmaceutical sector produces cost-effective, high-quality generics. A burgeoning bicycle industry aligns with eco-friendly transportation demands, while its expertise in jute offers sustainable solutions for textiles and packaging. The nation's home textiles, especially terry towels, and sports footwear industries are globally competitive. Additionally, Bangladesh can provide Latin America with affordable PVC bags, innovative furniture, ceramic tableware, acrylic yarn, and toys. Its robust leather industry and emerging motor parts sector further underscore its potential as a reliable trade partner for Latin American nations.

To fully realize these opportunities, it's essential for Bangladesh to address trade barriers such as tariffs, regulatory compliance, quality standards, and logistics challenges. Moreover, fostering diplomatic ties, conducting market research, and establishing reliable distribution channels will be crucial in entering and succeeding in the Latin American market. With concerted efforts and strategic partnerships, Bangladesh can undoubtedly tap into the vast potential that Latin America offers in these sectors.

About impacts of economic and trade relationships on bilateral relations with the other states she continued, the economic and trade relationships of Bangladesh play a crucial role in shaping its bilateral relations with other states in the region.

Regional Economic Integration: Bangladesh's economic ties with neighboring countries have a direct impact on regional economic integration. For example, its trade relations with India, Nepal, Bhutan, and Myanmar are vital components of the South Asian Association for Regional Cooperation (SAARC) framework. A strong and cooperative economic relationship with these nations can foster regional stability and development.

Trade Balance: The trade balance between Bangladesh and its neighbors is a significant factor. The country's exports, including garments, textiles, and agricultural products, have a substantial impact on its economic well-being. If Bangladesh can maintain a positive trade balance by exporting more than it imports, it can build stronger diplomatic ties with its neighbors through mutually beneficial trade agreements.

Infrastructure Development: Bilateral economic cooperation often involves infrastructure development projects like roads, railways, and ports. These initiatives can enhance connectivity and facilitate trade not only with immediate neighbors but also with other countries in the region. Well-executed infrastructure projects can boost Bangladesh's standing as a regional hub for trade and commerce.

Energy and Resources: Access to energy resources, such as natural gas and hydropower, can also influence Bangladesh's relations with neighboring countries. Collaborative efforts in the energy sector, such as cross-border pipelines or electricity-sharing agreements, can foster economic interdependence and promote peaceful relations.

Political Relations: Economic and trade ties are closely intertwined with political relations. When economic interests align, it often paves the way for improved diplomatic ties. Conversely, trade disputes or economic imbalances can strain bilateral relations. Bangladesh's ability to navigate these political challenges and maintain a balanced approach to economic cooperation is crucial.

Regional Stability: A strong and growing economy in Bangladesh contributes to regional stability. Economic development can reduce poverty, improve living standards, and enhance social cohesion. Stable neighboring countries are more likely to cooperate on security and regional stability issues, strengthening the overall fabric of regional relations.

Bangladesh's economic and trade relationships have a multi-faceted impact on its bilateral relations with neighboring states. These relationships can either foster cooperation, stability, and prosperity or lead to tensions and conflicts, depending on how they are managed and leveraged by the government of Bangladesh and its regional counterparts. Successful economic diplomacy and careful negotiation of trade agreements can play a pivotal role in ensuring positive outcomes for Bangladesh's regional relations.

In what ways may Bangladesh enhance its credibility in Latin American economic affairs by engaging in economic diplomacy within the global political framework? In her answer to this query, she states that in order to strengthen Bangladesh's standing in Latin American business dealings, economic diplomacy within the context of international politics necessitates a thorough plan that incorporates a range of stakeholders and makes use of available resources. Bangladesh can take the following actions and things into account to accomplish this goal:

If Bangladesh wants to increase its commercial and economic links to Latin America, it needs to conduct extensive market research to identify potential business partners and customer preferences in the region. It will be crucial to strengthen bilateral relations by holding high-level talks, forming trade commissions, and drafting trade agreements. Improved ties are possible if Bangladesh diversifies its exports beyond just textiles, agriculture, and technology and promotes its image abroad through advertising and cultural exchanges. It is essential to attract investments, organize trade missions, promote public–private partnerships, and stress the need of sustainable business practices. Diplomatic initiatives, combined with official capacity building, risk assessments, and constant monitoring and evaluations, will ensure the success of these endeavors.

By implementing a well-rounded economic diplomacy strategy that incorporates these elements, Bangladesh can strengthen its credibility and presence in Latin American economic activities. It will require persistence, adaptability, and a long-term commitment to building mutually beneficial relationships in the region.

Diplomat continued, "India, as a South Asian nation, has indeed made a significant impact and presence in South America over the years," when asked about the country's economic standing in the region. Bangladesh could benefit from India's dominant economic position in the area.

Diversification of Trade Partnerships: India has successfully expanded its trade relationships beyond its immediate neighbors, and this diversification has helped in reducing economic vulnerabilities. Bangladesh can also explore trade opportunities with South American countries to reduce dependency on a limited number of trading partners.

Economic Diplomacy: India's strong diplomatic and economic presence in South America has been partly driven by its proactive economic diplomacy efforts. Bangladesh can emulate this approach by actively engaging with South American nations through diplomatic channels to promote economic cooperation, investment, and trade agreements.

Investment in Infrastructure: India has invested in infrastructure development projects in South American countries, which has not only strengthened economic ties but also created goodwill. Bangladesh can consider similar infrastructure investments taking the advantage of the nearshoring that align with the development needs and priorities of South American nations, fostering stronger bilateral relationships.

Cultural and People-to-People Connections: India has leveraged its rich cultural heritage and diaspora communities in South America to build connections and facilitate economic interactions. Bangladesh can explore the promotion of cultural exchanges, tourism, and people-to-people connections to enhance its presence and influence in the region.

Knowledge Sharing and Capacity Building: India has often shared its expertise in various sectors, including information technology and agriculture, with South American countries. Bangladesh can follow suit by offering its knowledge and technical expertise to foster cooperation and mutual benefit.

Adaptability and Flexibility: India has demonstrated adaptability and flexibility in its approach to South America, tailoring its strategies to meet the specific needs and preferences of each country. Bangladesh can similarly customize its engagement strategies to accommodate the unique characteristics of South American nations.

Political Engagement: Like India, Bangladesh can consider a more active role in regional organizations and diplomatic initiatives to strengthen its presence and influence.

Market Research and Analysis: Conducting thorough market research and analysis is essential before entering South American markets. India's success in the region can be attributed, in part, to a deep understanding of the local market dynamics. Bangladesh can benefit from investing in such research to identify viable opportunities and challenges.

Sustainable Development: India has recognized the importance of sustainable development and environmental conservation in its engagements with South American nations. Bangladesh can adopt a sustainable approach to economic cooperation, focusing on environmentally responsible practices and initiatives.

Therefore, by diversifying trade partnerships, engaging in economic diplomacy, investing in infrastructure, fostering cultural ties, sharing knowledge, and maintaining adaptability and flexibility, Bangladesh can enhance its presence and economic influence in South America and forge stronger partnerships with countries in the region.

On Bangladesh's football diplomacy with Latin America, Mexican Ambassador pointed out that, Football, often referred to as the "beautiful game," possesses significant potential to enhance bilateral relations between Bangladesh and Latin American nations, as exemplified by Argentina. However, to realize this potential, Bangladesh must first establish a world-class football team. To unlock this potential:

Cultural Exchange: Football unites Bangladesh and Latin America, fostering cultural exchange and mutual appreciation through matches and tournaments.

Diplomatic Opportunities: Prestigious events like the FIFA World Cup provide diplomatic platforms for Bangladesh and Latin American countries to collaborate.

People-to-People Connections: Football fans form lasting friendships, boosting tourism and partnerships, enhancing trust.

Soft Power and Branding: Successful Latin American nations use football to enhance their international image, attracting investments and cultural exchanges. Bangladesh can follow suit.

Sports Diplomacy: Football supports diplomacy, addressing shared challenges and fostering cooperation.

Youth Engagement: Football engages youth and encourages collaboration among future generations.

Argentina's example illustrates how football catalyzes bilateral relations, promoting cultural exchange, diplomacy, personal ties, soft power, sports diplomacy, and youth engagement. Leveraging football's appeal can build stronger, interconnected relationships beyond the pitch.

With respect to Bangladesh's current diplomatic ties with South America, the Bangladeshi government's Assistant Secretary of International Trade, Investment & Technology Wing, MoFA, stated that Bangladesh is currently exhibiting a strong desire to establish relationships with the countries in this region. As a result, diplomatic relations in that area are already trending higher. Although in the past, Bangladesh was unable to give opening an embassy or mission their top priority. Nevertheless, ongoing endeavors are being made to address this issue. In a recent development, the distinguished State Minister embarked on an official visit to Brazil, marking the first occasion of such a diplomatic endeavor aimed at commemorating the fifty-year milestone of the bilateral relationship between Brazil and Bangladesh. Additionally, it is noteworthy that our nation's foreign secretary has undertaken official visits to the countries of Chile and Brazil. Significant advancements are now underway in Chile regarding the evaluation of diverse trade agreements and other pertinent matters. In addition to this, efforts are currently being made to initiate the process of re-establishing the diplomatic mission of Bangladesh in Argentina.

The Assistant Secretary says, "In the immediate aftermath of Bangladesh's independence, several Latin American nations recognized the country." This answer addresses the topic of how the economic ties between Bangladesh and the Latin area have grown over the past 50 years. Still, the bond did not develop quickly. The fundamental basis of this issue may be traced back to the disparities that exist within our society, culture, and political landscape. Furthermore, the level of interpersonal communication among the general population is significantly limited. However, as a result of economic requirements, Bangladesh relies on the markets of Brazil and neighboring nations for the procurement of goods. Presently, Bangladesh engages in substantial imports from Brazil. The bilateral communication between Bangladesh and the respective country is seeing an upward trajectory. It is plausible that there may be a potential for enhanced economic relations with other nations, such as Chile and Argentina, in the forthcoming period.

About challenges to strengthening the economic relationship with the region's different countries, Assistant Secretary suggests that, Distance does not serve as a determinant for international trade, but rather acts as a regulatory component. Nations characterized by larger geographical spans often share alternate supply chain networks. Which is applicable to both Bangladesh and countries in the Latin region.

Additionally, one must consider the rigid customs and practices exhibited by the Latin population. They have distinct laws than other parts of the world. As a result, the area's commercial and tourism potential are at odds.

Moreover, the visa application procedure for Latin American countries poses significant complexities for individuals from Bangladesh. As a result, the pace of rise in two-way communication among the general population is not significant. To enhance the economic ties between two nations, it is imperative for the corporate communities of both countries to conduct comprehensive research about consumers and marketplaces. The enhancement of economic interactions necessitates the presence of bilateral demand. Both consumers and traders are required to produce it. The enhancement of cultural convergence is anticipated to have a positive impact on economic connections. Nevertheless, the growth of economic contacts is not meeting expectations as a result of procedural errors, bureaucratic misuse, and corruption occurring on both sides.

Discussing roles of public–private partnerships in relationship she gives insights, Assistant Secretary explained that, the level of international trade in Latin American countries varies from state to state. It is important to engage in diplomatic endeavors at the governmental level in order to enhance the magnitude of this commerce. If the government can establish a diplomatic position, it will stimulate private organizations and traders to engage in import and export activities. When corporate delegations attempt to develop their commerce through official means, their acceptance internationally rises. In order to address this matter, it is imperative that business organizations take proactive measures. The significance of inter-ministerial collaboration at the state level cannot be overstated in this context. The government of Bangladesh is now engaged in addressing these concerns. Business delegations have been dispatched to Latin America during various official visits.

The Latin area has the potential to emerge as a viable alternative to Bangladesh's export market. Given the existing valuation of $1 billion, there exists potential to augment it to $5 or $10 billion. The trade dependency of Bangladesh on certain nations is expected to diminish progressively. It will assist Bangladesh in establishing a robust economy and rising to the status of a high-income nation.

The removal of supplementary tariffs on Bangladeshi goods inside the Latin region would enhance the country's ability to effectively compete with other nations. However, despite the potential of Bangladesh to export trained labor to Latin American countries, this prospect is hindered by linguistic obstacles. Hence, Bangladeshis endeavor to migrate to the United States, even if they relocate to these aforementioned areas. The potential of Bangladeshi products in Latin America is substantial, contingent upon their ability to navigate and overcome challenges within the international market.

About impacts of economic and trade relationships on bilateral relations with the other states she continued Assistant Secretary briefly said, the establishment of commercial and economic ties naturally facilitates the progression of bilateral relations between two countries. Bangladesh exhibits limited trade engagement with Latin American nations, with the exception of Brazil. Bangladesh considers these nations as non-traditional markets. Nevertheless, the government of Bangladesh periodically sent delegations to various nations in order to assess the trade prospects. In such a scenario, the attainment of diplomatic success would likely result in a further augmentation of the desire for bilateral relations. Consequently, there would be a progression in international bilateral relationships.

In what ways may Bangladesh enhance its credibility in Latin American economic affairs by engaging in economic diplomacy within the global political framework? A junior diplomat responded to this query by saying that Bangladesh has already started working on economic diplomacy. The international trade wing of our ministry has achieved significant progress in the field of economic diplomacy. In this particular instance, it is necessary to engage individuals who possess advanced expertise and proficiency in the field of economic diplomacy. Previously, Bangladesh did not accord major importance to the domain of economic diplomacy. Currently, there has been a slight improvement in the situation. The Ministry of Foreign Affairs orchestrated a seven-day specialized campaign on the subject of economic diplomacy in the year 2022.

The prioritization of public diplomacy over economic diplomacy is essential. Bangladesh should engage in public diplomacy efforts to establish its brand presence in the Latin region. The creation of an appealing reputation for Bangladesh has the potential to lead to a rise in demand for products originating from the country. Then, economic diplomacy will be successful in the Latin American region. Trade partnerships exist with countries that possess significant global influence. Therefore, it is vital for Bangladesh to strive for survival through competitive means. Even though China and the United States are attempting to preserve their commercial hegemony in the Latin American region, Bangladesh will not be threatened or hampered by this. The potential for generating demand through product quality and production exists.

When asking in the context of India's economic standing in South America, he said, drawing a comparison between India and Bangladesh may not be seen as a realistic endeavor. The efforts to build the Indian market in the Latin region were undertaken in the past. Moreover, a significant proportion of the Indian population resides within the Latin region. Indigenous populations, particularly in nations such as Suriname and Venezuela, wield significant political influence. Bangladesh is required to address matters pertaining to the Latin area autonomously. To accomplish this objective, it is crucial to enhance diplomatic ties through the establishment of diplomatic missions in the respective nations.

On Bangladesh's football diplomacy with Latin America, he remarks, Countries in Latin America are at the forefront of global football. In contrast, Bangladesh is positioned at the lower end of the spectrum in terms of its performance in football. Even so, the endorsement of Messi, Pele, and Maradona by the Bangladeshi population garnered significant attention from the citizens of that specific region. It is advantageous from a diplomatic standpoint. This will result in an increase in bilateral communication among ordinary individuals. The surge in football popularity has led Argentina to establish a diplomatic mission in Bangladesh. It is indisputable that this assertion holds true. However, the implementation of football diplomacy may not be deemed as a viable and enduring programmer or tactic. The achievement of significant progress is unlikely without the establishment of shared economic, political, and cultural interests. Therefore, it can be asserted that the embassies of Argentina and Brazil in Bangladesh are not operating at its entire capacity. The achievement of this objective necessitates the identification and fulfillment of shared interests, as well as the establishment and maintenance of appropriate diplomatic ties.

6.1.1 Relevant Personalities' Thoughts on Latin America and Economic Diplomacy

Officials at all levels in Bangladesh's government have recently sounded upbeat about the country's potential for doing business with Latin America. Since it was not possible to conduct interviews with the aforementioned people for this study, the relevant viewpoints have been gathered from various mainstream media sources. These perspectives have contributed significantly to the investigation and analysis of empirical data.

The imperative of boosting trade and business collaboration between Brazil and Bangladesh was emphasized by the prime minister of Bangladesh. The potential for business growth between the two nations was enormous. The government of Brazil has been requested by Prime Minister Sheikh Hasina to prioritize the growth of bilateral commerce and business. As Bangladesh would become a developing country in 2026, she underlined the importance of signing preferential trade agreements or free trade agreements with Brazil and the other three members of MERCOSUR (The Southern Common Market): Argentina, Paraguay, and Uruguay. The PM requested that the accords be signed as quickly as possible, noting that talks for a PTA or FTA with the MERCOSUR countries are still ongoing. Bangladesh primarily imports soybeans, raw sugar, and pharmaceuticals from Brazil as well as RMG goods.

Bangladeshi diplomats have been instructed by Prime Minister Sheikh Hasina to take a more proactive role in bolstering economic diplomacy, with the goal of preserving the country's status as a developing country by boosting export and commerce. "You (diplomats) will have to be active in strengthening economic diplomacy alongside brightening the image of the country," she said on Monday night at a regional envoy conference held at her home here. Diplomats from Bangladesh serving in Middle Eastern nations became members of it. She stated that since Bangladesh would soon become a developed nation, focus should be on those nations where Bangladesh can grow its trade and business and gain advantages as a developing nation.

The prime minister added that in order to understand and sustain product demand and quality, research is necessary for trade and commerce to grow. In order to increase exports and company growth, she advised the nation's businessmen to focus on research for product diversity and own branding. "Diplomacy has become virtually exclusively economic in today's world. Our diplomatic missions' responsibilities have likewise evolved. The prime minister stated, "It's more like economic diplomacy now, not just political diplomacy." At this point, we must consider how to increase commerce and trade, attain economic independence through upholding amicable relations with all parties, accomplish progress, and foster peace by working together. That's how we'll have to conduct diplomacy," she remarked.

In an article titled "Economic diplomacy: A conduit for Bangladesh's vision 2041," the former foreign minister of Bangladesh, Abdul Momen, asked, "What is Economic Diplomacy?" The five elements of the economic diplomacy package are as follows: (1) increased foreign investment; (2) increased trade and export diversity; (3) gainful employment of human resources both domestically and internationally; (4) technology transfer; and (5) quality services to the Bangladeshi diaspora and other customers.

In a statement published in the Daily Star, the government of Bangladesh's then-foreign minister, Dr. AK Abdul Momen, stated that the country is currently concentrating on economic diplomacy in an effort to attract firms and grow its economy. The minister stated, "Bangladesh is now focused on economic diplomacy to connect to more countries and to find new avenues of bringing businesses to Bangladesh," during a press conference this afternoon at the Sylhet Circuit House.

Then-foreign minister stated that Bangladesh aspires to become a full member of Mercosur, the formidable South American economic bloc that also includes Uruguay, Brazil, Paraguay, and Argentina. "We have filed our petition to become a member of Mercosur, but they (Argentina) are currently the organization's president. Our trade in Latin America would increase significantly if we were a member of Mercosur," he said following a meeting with Santiago Andres Cafiero, the Argentine Minister of Foreign Affairs, International Trade, and Worship.

According to the foreign minister, he asked his Argentinean counterpart to help Bangladesh join Mercosur. Momen stated that if Dhaka is able to handle its membership in that trade alliance, which also includes Bolivia, Chile, Colombia, Ecuador, Guyana, Peru, and Suriname, then Bangladeshi products would encounter less tariff barriers in the South American market.

According to him, Bangladesh sought to acquire edible oil from Argentina, while Dhaka pushed Buenos Aires to import more goods created in Bangladesh, such as medicines and RMG. According to Momen, they also talked about ways to promote South-South collaboration between the two nations.

For our medications, leather goods, ceramic wares, plastic and melamine products, IT goods and services, jute goods and handicrafts, and agricultural goods and services, we are targeting the sizable Latin American market. Bangladesh desires to investigate new avenues for collaboration with all of the surrounding nations. Bangladesh's exports of clothing to Latin American nations have increased steadily during the past few years. He stated during a situation.

The then Foreign Secretary, Bangladesh Government had made a rapid tour of Latin America and saw Bangladesh's immense business potential in countries like Brazil to Argentina and Panama to Mexico. Since Bangladesh has no residence mission in any of the Latin American countries at present, one of the objectives of the mission was to establish direct contacts with those countries to diversify the country's markets beyond North America and Europe. Earlier, Bangladesh had missions in both Brazil and Argentina, but those were wrapped up at later stage. Now, the government has initiated process to set up a full-fledged mission in Brazil, an emerging economy in Latin America.

"We've deep engagements with the countries of Latin America in multilateral forums, and now we want to transform those engagements into bilateral cooperation in expanding trade and investment," he said. The team—comprising officials from the ministries of commerce, overseas employment, agriculture and foreign affairs—visited Peru, Chile, Brazil, Argentina, Mexico, Panama, Jamaica and Colombia from July 28 to August 10.

Referring to the prospects of Bangladeshi products, the foreign secretary said Bangladeshi apparels, jute products, pharmaceuticals and handicrafts have potentials in those countries. Volume of trade between Bangladesh and the Latin American countries is insignificant, may be in thousands of dollars. During the interactions with officials and businessmen there, the secretary said they have discussed framework agreements, technical cooperation, particularly in rice and maize production. In Chile the delegation discussed visa waiver agreement for Bangladeshi diplomats and government officials. The negotiations with Brazil regarding the visa waver have been completed. He said Mexico, Colombia and Argentina have also expressed their willingness to sign visa waiver agreement.

Gradually, he said, this programmer would be expanded for Bangladeshi businessmen. He also said Argentina, Mexico and Colombo expressed their interests in conducting Spanish language course for Bangladesh officials. The secretary also sees prospects of sending Bangladesh engineers and construction workers to Panama as the Panama government is expanding the canal to facilitate the sailing of big merchandise ships through it.

Brazilian ambassador to Bangladesh mentioned that, Bangladeshi readymade garments had a great demand in the Brazilian market. Brazil can be a major market for Bangladeshi RMG products. There was a huge scope to strengthen the bilateral cooperation between the two countries, particularly in trade and technology. He emphasizes on enhancing the people-to-people and business-to-business contact between the two countries. Regarding the bilateral agricultural cooperation agricultural good practices, experiences and technology can be shared with each other.

Mentioning that trade is very concentrated in garments and commodities, he said: "I am confident that with time we can diversify it. The two-way trade stood at $2 billion without institutional help." "So, we believe it is high time to give it a push," he added. The ambassador observed that both politically and economically, Bangladesh is receiving much more attention from Brasilia. The country is becoming better known among the Brazilian public who wear more and more "Made in Bangladesh" garments.

He mentioned that during their meeting on the sidelines of the BRICS summit in Johannesburg, President Lulu and Prime Minister Sheikh Hasina agreed to improve cooperation on trade, livestock, agribusiness, energy, infrastructure, and ICT, among other sectors. "They will probably meet again during the upcoming G20 summit in New Delhi," he said, as Brazil is a member of G20 and Prime Minister Sheikh

Hasina will join as an invited guest. The ambassador remarked that President Lulu also showed interest in visiting Bangladesh. "I hope he will visit in future, which would be great for bringing the two countries closer together and enhancing our trade and cooperation," he said.

In the Daily Star, the Mexican Ambassador to Bangladesh was reported as saying, "In the past five years, the two-way trade multiplied by 100 percent to about $500 million, and Mexico is now eager to import pharmaceutical and IT products." In order to diversify its export portfolio, Bangladesh is actively looking to develop new markets, particularly for pharmaceuticals, IT services, and leather goods, according to the Director General of the South America Wing of the MoFA, Bangladesh. Thus far, South America has been virtually unexplored. Thus, we hope to strengthen our ties with the nations of Latin America.

This is important as Bangladesh graduates out of LDC by 2026, when some trade privileges may be curtailed. Also, with the disruptions of supply chains following the Russia-Ukraine war, all countries are looking for diversified sources and markets. Bangladesh is no exception. Similarly, countries like Argentina and Mexico are also eager to boost relations with us. Bangladesh is also a large and growing market.

Bangladesh lags behind in business in Latin America. Many products are imported into Bangladesh via different countries. As a result, products' prices go up here, he added. "So, we need to establish direct communications with Latin American countries to get our required products at reasonable prices. However, we hope that our upcoming networking summits will help expand our business and increase export volume as well." He also explained that Bangladesh's dependency on crude oil will come down if the country can import directly from Latin America, particularly from Venezuela instead of Middle East countries. "Bangladesh has to import poultry and fish feeds, edible oil, wheat, and construction materials like scraps from the Latin American countries. But we have to go to Latin America via India, Dubai or EU which creates hassle for us," he pointed out. "Now, a ship needs 20 days to reach Chattogram port from Latin America after landing two /three ports of Dubai, Singapore or EU. So, it puts an impact on carrying costs of products. Besides, air transportation system is also the same. If we can communicate directly among the countries, it will boost our business definitely."

Chapter 7
Bangladesh and LAC: Issues of Bilateral Relations and Prospects

Abstract This chapter examines the complexities and prospects of bilateral relations between Bangladesh and Latin American countries (LAC), focusing on economic and strategic dimensions. With a geographical separation exceeding 10,000 miles and limited historical and cultural ties, the chapter highlights the challenges and opportunities shaping these relations in an evolving global landscape. Despite the significant geographical distance—ranging from 8000 to nearly 10,000 nautical miles—technological advancements and global trade dynamics reduce its deterrent impact on economic ties. Comparisons with Bangladesh's existing trade relationships with distant partners like the United States and Brazil suggest that logistical challenges can be mitigated with proper infrastructure and strategic policies. The chapter underscores the critical need for diversified markets, emphasizing LAC's potential to complement Bangladesh's import and export demands. Bangladesh's exports—dominated by textiles, jute, pharmaceuticals, and ceramics—align with LAC's demand for such products. Conversely, LAC's primary exports, including soybeans, crude petroleum, minerals, and agricultural goods, align with Bangladesh's import needs. The analysis illustrates significant trade potential, particularly with Brazil, whose export profile complements Bangladesh's economic requirements. The establishment of direct trade channels and preferential agreements can unlock these synergies, as evidenced by LAC's rising trade with competitors like India and China. Multiple barriers impede deeper bilateral engagement, including linguistic and cultural differences, inadequate direct transportation links, and minimal historical or political ties. Intra-ministerial coordination gaps and information asymmetry further complicate trade initiatives. The absence of bilateral or multilateral trade agreements results in regulatory hurdles and limited competitive advantage in LAC markets. The chapter also emphasizes the need to address logistical inefficiencies, such as extended shipping times and high transportation costs. Contemporary global shifts, including supply chain diversification, South-South cooperation, and climate action imperatives, create a conducive environment for Bangladesh-LAC partnerships. Joint ventures in renewable energy, sustainable agriculture, and infrastructure development offer promising avenues for collaboration. The chapter suggests leveraging digitalization and fostering public–private partnerships to facilitate economic

M. T. Islam and M. B. Hossen, *Economic Diplomacy*, The Latin American Studies Book Series, https://doi.org/10.1007/978-3-031-81362-7_7

ties. The proposed model offers a comprehensive approach to deepening bilateral relations by addressing existing challenges and leveraging shared economic and strategic interests. By implementing this framework, Bangladesh and LAC can achieve sustainable, mutually beneficial growth and enhance their global economic standing. The proposed framework integrates free trade agreements, public–private partnerships, and enhanced connectivity to mitigate logistical constraints. It also underscores the role of academic and cultural exchanges in bridging linguistic and cultural divides, alongside digital platforms and virtual trade fairs to overcome physical distances. This analysis identifies LAC as a pivotal market for Bangladesh's trade diversification strategy. Overcoming structural and cultural challenges through targeted policies, robust economic diplomacy, and strategic infrastructure investments is imperative. By aligning with global trends and fostering mutual understanding, Bangladesh and LAC can establish a dynamic and resilient economic partnership, driving long-term growth and development for both regions.

Keywords Economic diplomacy · Bangladesh · LAC · Bilateral relations · South-South cooperation · International trade

7.1 Is Geographic Distance the Most Significant Barrier to the Growth of Connections?

The geographical location and distance between two countries play a significant role in shaping the dynamics of their international relations. The ability to establish and foster diplomatic relations with geographically distant nations is a feat that can only be achieved by powers with enormous power and influence. For instance, countries such as the United States, United Kingdom, China, and Russia. However, emerging nations such as Bangladesh primarily engage in diplomatic ties with neighboring countries and global powers. In contemporary times, the global community has experienced a major transformation into a closely interconnected entity, primarily attributed to advancements in technology. The concept of the global village was established by Marshall McLuhan because to this explanation. Hence, in the contemporary day, nations across the globe are forging bilateral partnerships with other nations in accordance with their economic and political exigencies. Consequently, there is currently a growing phenomenon known as the emergence of the 'Global South' taking place worldwide.

The mean geographical separation between Bangladesh and countries in Latin America exceeds 10,000 miles. Moreover, it is noteworthy that the mean distance traversed by sea amounts to approximately 9000 nautical miles. There exists a notable absence of linguistic, cultural, and lifestyle similarities between Bangladesh and the aforementioned region. The colonial legacy of Bangladesh does not have any direct relevance to this particular location. Despite the absence of historical connections, economic requirements can facilitate the integration of two countries and domains.

As it can be said of Bangladesh and Brazil's relationship. In the contemporary global context, geographical distance no longer serves as a hindrance to the continuous flow of international relationships and trade. The consolidation of the global commerce system has been facilitated by advancements in transportation infrastructure and technology. China and India serve as exemplary illustrations in this context. Currently, the two adjacent nations of Bangladesh serve as the primary economic allies within the Latin American region.

The primary hindrance to the establishment of connections between the Latin area and Bangladesh is the enormous geographical distance between the two regions. Through careful analysis and observation of the given situation, one can gain a comprehensive understanding of the underlying complexity inherent in this inquiry. The formation of relationships with a particular region is influenced by various logical variables, in addition to geographical location.

These are

1. Cultural and socioeconomic differences between Latinos and Bengalis.
2. There hasn't been any established interactions between the populations of the two regions.
3. There are very few bilateral tourism connections. The pricey transport system is the main cause of this.
4. There is no colonial legacy in Bangladesh or Latin America. While Spain, Portugal, Italy, and France governed the region, the British ruled Bangladesh.
5. The global political context and policy-making organizations are also indifferent.
6. The absence of robust historical ties.
7. There aren't any political or religious expectations.

The aforementioned elements are hindering the growth of Bangladesh's relations with Latin America on a global scale. Modern technology makes it possible to eliminate these consequences. Additionally, locating substitute markets to satisfy economic demands will aid in overcoming this disadvantage. To achieve this, the country's internal and foreign policies need to undergo the appropriate modifications and additions.

7.2 An Analysis of the Correlation Between Economic Activity and Geographical Location

Distance analysis between Bangladesh and 13 Latin American countries is required to grasp the fundamental fact of this study. Distances between countries in South America range from 8123 to 9653 nautical miles, with Chile being the farthest from Bangladesh. Additionally, Bangladesh is located 9053 nm from Argentina and 8414 nm from Brazil (Table 7.1).

Meanwhile, the average distance to Bangladesh's top 10 importing countries is over 2000 nm. China and India are Bangladesh's top two suppliers. A difference of 1634 statute miles and 769 nautical miles. Within 2000 miles, Bangladesh also

Table 7.1 Distance between Bangladesh and Latin American Countries

Country name	Distance in kilometers (km)	Distance in miles (miles)	Distance in nautical miles (nmi)
Argentina	16,767.64	10,418.93	9053.80
Brazil	15,582.98	9682.81	8414.13
Bolivia	17,638.70	10,960.18	9524.14
Chile	17,879.12	11,109.57	9653.96
Colombia	16,445.62	10,218.84	8879.93
Ecuador	17,142.66	10,651.95	9256.30
Venezuela	15,482.81	9620.57	8360.05
Paraguay	16,784.63	10,429.49	9062.98
Peru	18,162.44	11,285.61	9806.93
Suriname	15,044.34	9348.12	8123.30
Uruguay	16,565.42	10,293.28	8944.61
Guyana	15,203.89	9447.26	8209.44
French Guiana	14,872.38	9241.27	8030.44

Source Distancefromto website

receives imports from Singapore, Indonesia, Japan, Malaysia, Korea, and the United Arab Emirates. Bangladesh, on the other hand, imports from countries with the longest shipping times to seventh- and fourth-ranked Brazil and the United States, respectively. A total of 8414 statute miles and 6982 nautical miles separates the two nations. In other words, it is possible to import it from faraway places if there is a need or desire. If the necessary infrastructure is in place, this situation lends itself to increased imports and exports (Table 7.2).

Table 7.2 Geographical distance of top import partners of Bangladesh

Country name	Distance in kilometers (km)	Distance in miles (miles)	Distance in nautical miles (nmi)
China	3026.43	1880.54	1634.14
India	1425.14	885.54	769.51
Singapore	2882.56	1791.14	1556.46
United States	12,932.48	8035.87	6982.98
Japan	4894.21	3041.12	2642.66
Indonesia	4894.21	3041.12	2642.66
Brazil	15,582.98	9682.81	8414.13
Malaysia	2587.93	1608.06	1397.37
UAE	3650.88	2268.55	1971.32
Korea	3730.23	2317.86	2014.16

Source Distancefromto website and Business Inspection BD

Table 7.3 Top export partners of Bangladesh: geographical distance

Country name	Distance in kilometers (km)	Distance in miles (miles)	Distance in nautical miles (nmi)
United States	12,932.48	8035.87	6982.98
Germany	7073.96	4395.55	3819.63
United Kingdom	8004.34	4973.66	4322.00
Spain	8655.41	5378.22	4673.55
France	7911.28	4915.84	4271.75
Poland	6560.25	4076.35	3542.25
Italy	7308.46	4541.27	3946.25
India	1425.14	885.54	769.51
Netherlands	7647.28	4751.80	4129.20
Canada	12,200.74	7581.19	6587.87

Source Distancefromto website and Business Inspection BD

There is room for growth in bilateral relations as well as the economic collaboration with the countries. However, Western Europe and North America account for nine of Bangladesh's top ten export markets. Distances between Bangladeshi ports and the top ten nations average over 5000 nautical miles. India is the lone non-conformist on this list. That is, if Bangladesh can develop its own consumer market, distance won't be a barrier. Instead, the growth of these economic ties is strengthening Bangladesh's international standing. Bangladesh must develop its own potential in every region of the world by strengthening economic links if it is to advance from an emerging economy to a developed one. Because of this, it is important to use practical and effective economic diplomacy to counteract the negative effects of location (Table 7.3).

7.3 Economic Demands: Key Factors Affecting Growing Bilateral Relationships

The economic relationship between Latin America and Bangladesh has great potential for growth. In this regard, it is instructive to compare Bangladesh's imports with those of the region's exports. Cotton, machinery, mineral fuels and oils, electrical machinery and electronics, iron and steel, plastics, vehicles, man-made staple fibers, animal or vegetable fats and oils, and cereals are among the top imports into Bangladesh. Soybeans, crude petroleum oils and oils obtained from bituminous minerals, iron ores and concentrates, waste petroleum oils and oils obtained from bituminous minerals, maize (corn), cane sugar, beetroot sugar, chemically pure sucrose, bovine meat, coffee, etc. are the top ten exports from Latin America (Table 7.4).

Table 7.4 Top import countries of Bangladesh by percentage

Country	Value USD %
China	21
India	12
Singapore	9
Hong Kong	5
Indonesia	4
Japan	3
South Korea	3
Malaysia	3
Brazil	2
Kuwait	2

Source Foreign Trade Statistics 2021–22 by BANGLADESH BUREAU OF STATISTICS (BBS)

As a result, Latin American countries can serve as a secondary market for a variety of Bangladeshi imports, including but not limited to mineral fuels and oils, iron and steel, animal and vegetable fats and oils, cereals, maize (corn), artificial filaments, sugar and sugar confections. Imports from Brazil, a country in South America, have already begun into Bangladesh. Presently, 2.7% of all imports into the country come from Bangladesh (Table 7.5).

In which case China contributes 21%. If Bangladesh is interested, it can lessen its reliance on specific countries' imports by engaging in trade with Latin American countries. Some of Bangladesh's most important imports can be found among the products exported by Brazil, the region's dominant trading power.

Table 7.5 Top import products of Bangladesh by percentage

Product	Value USD %
Cotton	12
Machinery	11
Mineral fuels and oils	8
Electrical machinery and electronics	7
Iron and steel	4
Plastics	4
Vehicles	3
Man-made staple fibers	3
Animal or vegetable fats and oils	3
Cereals	3

Source Foreign Trade Statistics 2021–22 by BANGLADESH BUREAU OF STATISTICS (BBS)

7.3.1 *Brazil's Top Exports in 2022: According to Trend Economy*

1. **13.9%** ($46 billion): Soya beans, whether or not broken
2. **12.7%** ($42 billion): Petroleum oils and oils obtained from bituminous minerals, crude
3. **8.63%** ($28 billion): Iron ores and concentrates, including roasted iron pyrites
4. **3.89%** ($13 billion): Petroleum oils and oils from bituminous minerals, not crude; preparations, i.e., containing by weight 70% or more of petroleum oils or oils from bituminous minerals; these being the basic constituents of the preparations; waste oils
5. **3.66%** ($12.2 billion): Maize (corn)
6. **3.28%** ($11 billion): Cane or beet sugar and chemically pure sucrose, in solid form
7. **3.27%** ($10.9 billion): Meat of bovine animals, frozen
8. **3.09%** ($10.3 billion): Oil-cake and other solid residues, whether or not ground or in the form of pellets, resulting from the extraction of soybean oil
9. **2.65%** ($8.88 billion): Meat and edible offal, of the poultry of heading 01.05, fresh, chilled or frozen
10. **2.55%** ($8.54 billion): Coffee, whether or not roasted or decaffeinated; coffee husks and skins; coffee substitutes containing coffee in any proportion

Source World Bank (2023)

Once again, goods from Latin America that are imported to Bangladesh are quite comparable to those exported from Bangladesh. Which is readily apparent from comparing the two countries' export and import tally (Table 7.6).

Products that Bangladesh typically ships to the aforementioned nations also sell well in Latin America. Even so, Bangladesh's ties to Latin America were too immature for the country to flourish there. However, China and India are both launching proper economic diplomacy to compete in the Latin American market from around

Table 7.6 Top export countries of Bangladesh

Export partner	Value ($ in millions)	% of the total
USA	10,417.72	20
Germany	7590.97	14.57
United Kingdom	4828.08	9.27
Spain	3166.37	6.08
France	2711.06	5.21
Netherland	1775.01	3.41
Italy	1702.29	3.27
Canada	1522.96	2.92
Japan	1353.85	2.60
Belgium	900.03	1.73

Source Foreign Trade Statistics 2021–22 by BANGLADESH BUREAU OF STATISTICS (BBS)

Table 7.7 Top export products of Bangladesh

Product	Total export in %
Readymade garments	81.82
Jute and jute manufacture	2.17
Fish, shrimps, and prawns	1.02
Leather and leather manufactures	2.39
Home textile	3.11
Vegetable products	2.23
Pharmaceutical products	0.4
Plastic and plastic products	0.3
Bicycle	0.3
Terry towel	0.1

Source Foreign Trade Statistics 2021–22 by BANGLADESH BUREAU OF STATISTICS (BBS)

the same distance as Bangladesh. While Brazil, Argentina, and Bolivia now rely on China more than any other country for trade (Table 7.7).

A comparative picture of Bangladesh's main exports and Brazil's main imports:

7.3.2 Brazil's Top Imports in 2022: According to Trend Economy

1. **8.44%** ($24 billion): Petroleum oils and oils from bituminous minerals, not crude; preparations n.e.c. containing by weight 70% or more of petroleum oils or oils from bituminous minerals; these being the basic constituents of the preparations; waste oils
2. **3.47%** ($10.1 billion): Petroleum oils and oils obtained from bituminous minerals, crude
3. **3.28%** ($9.6 billion): Mineral or chemical fertilizers, potassic.
4. **2.81%** ($8.23 billion): Mineral or chemical fertilizers containing two or three of the fertilizing elements nitrogen, phosphorus and potassium; other fertilizers; goods of this Chapter in tablets or similar forms or in packages of a gross weight not exceeding 10 kg
5. **2.8%** ($8.2 billion): Parts and accessories of the motor vehicles of headings 87.01 to 87.05
6. **2.5%** ($7.32 billion): Mineral or chemical fertilizers, nitrogenous
7. **2.42%** ($7.09 billion): Insecticides, rodenticides, fungicides, herbicides, anti-sprouting products and plant-growth regulators, disinfectants and similar products, put up in forms or pickings for retail sale or as preparations or articles (for example, Sulfur-treated bands, wicks and candles, and fly-papers)
8. **2.26%** ($6.62 billion): Petroleum gases and other gaseous hydrocarbons
9. **2.1%** ($6.16 billion): Semiconductor devices (e.g., diodes, transistors, semiconductor based transducers); including photovoltaic cells assembled or not in modules or panels, light-emitting diodes (LED) assembled with other LEDs or not, mounted piezo-electric crystals

(continued)

(continued)

10. **2.06%** ($6.02 billion): Coal; briquettes, ovoid's and similar solid fuels manufactured from coal

Source World Bank (2023)

There are two primary economic drivers in Latin America: agriculture and mining. Minerals and other raw materials abound in wide swaths of land in Latin America. Furthermore, Latin America's tropical and temperate temperatures are perfect for cultivating a wide range of food crops. Agricultural goods and natural resources like copper, iron, and petroleum make up the bulk of Latin America's exports. Mineral fuels, including oil, machinery, including computers, electrical machinery, equipment, fertilizer, organic chemicals, vehicles, plastics, plastic articles, pharmaceuticals, other chemical goods, optical, technical, medical gear, and so on and so forth are the most common imports.

Taking everything into account, it seems reasonable to conclude that the economic demands of Bangladesh and Latin America are complementary, despite the fact that there are cultural and linguistic differences between the two regions. To meet these economic requirements through import–export or international trade, a thriving international community is essential. India has already taken a number of steps to expand its commercial ties to the region of Latin America. As a result, by 2022, transactions totaling roughly $50 billion will have taken place. By expanding bilateral ties with 13 nations in Latin America, India has achieved a level of success never before seen. Bangladesh's foreign policy priorities should reflect this economic necessity by giving high priority to expanding ties with Latin America. The leaders of these 13 nations have not paid a state visit to Bangladesh yet, which is unfortunate.

Prior to this, in 2011, Mirazul Kayes, then the Foreign Secretary of Bangladesh, visited many Latin American nations. After that date, no further details of a state visit by a Bangladeshi delegation have surfaced. The group traveled to Peru, Chile, Brazil, Argentina, Mexico, Panama, Jamaica, and Colombia between July 28 and August 10, consisting of officials from the ministries of commerce, overseas employment, agriculture, and foreign affairs. Meanwhile, Argentine Foreign Minister Santiago Andrés Cafiero has confirmed that he will travel to Bangladesh in February 2023. Bangladesh has already come to terms with its economic situation and has set its sights on Latin America. Whose actions have been particularly outstanding. In January of 2023, for instance, Bangladesh's Ministry of Foreign Affairs set up a Latin America Department. The Daily Star and other Bangladeshi media outlets stated in February 2023 that, in the wake of the Russia-Ukraine war's disruption of global supply networks, Bangladesh was trying to establish diplomatic and trade connections with Latin American countries.

Mozammel Haque, director general of the foreign ministry's South America department, stated in a media report that Bangladesh is actively striving to expand exports into new markets, particularly for pharmaceuticals, information technology services, and leather goods. South America is largely undiscovered, according to Mozammel Haque. Thus, we seek to expand our ties to the nations of Latin America. By 2026, Bangladesh will no longer be considered an LDC, and as a result, it may lose certain trade privileges. As a result of supply chain interruptions caused by the conflict between Russia and Ukraine, nations worldwide are searching for alternative markets and suppliers. The same holds true for Bangladesh.

7.4 Intra-ministerial Barriers for International Trade: Bangladesh-Latin American Countries Perspective

The complex mosaic of international trade often extends beyond tariffs, quotas, and geographic distances. One overlooked dimension is the intra-ministerial barriers, which refer to obstacles arising from within a country's governmental apparatus, affecting its external trade relations. From the perspective of Bangladesh and Latin American countries, several intra-ministerial barriers can be identified.

Lack of Inter-Ministerial Coordination

Efficient trade requires coordinated efforts from various ministries like commerce, finance, transportation, and foreign affairs. In the context of Bangladesh and some Latin American countries, there's sometimes a lack of synchronization among these ministries leading to conflicting policies and bureaucratic delays.

Ambiguous Policy Directives

Unclear and frequently changing policy directives from different ministries can create confusion for exporters and importers. For instance, a product may be identified as a priority export item by the commerce ministry, but financial incentives or logistical support might not align with this priority (*Interview 1*).

Inefficient Customs and Regulatory Procedures

Customs departments and regulatory bodies, often operating under different ministries, might not be streamlined, leading to tedious and prolonged clearance processes. This deters efficient trade transactions and elevates transaction costs (*Interview 3*).

Inconsistent Quality Standards

Ministries responsible for setting quality and safety standards might not always align their criteria with international benchmarks or with those of their trading partners. Such discrepancies can pose barriers to exporting products that don't meet the importing country's standards (*Interview 1*).

Limited Infrastructure Development

While transportation and infrastructure generally fall under specific ministries, their priorities might not always align with trade objectives. Inadequate infrastructure can impede the smooth flow of goods, especially for countries like Bangladesh with logistical challenges (*Interview 2*).

Inadequate Representation in Trade Negotiations

During trade negotiations, the absence of representatives from pertinent ministries can lead to agreements that don't account for holistic national interests. A trade deal might favor one sector but inadvertently harm another due to lack of comprehensive representation (*Interview 2*).

Lack of Information Dissemination

There's often a noticeable gap in information dissemination between ministries and businesses. Local businesses might remain unaware of trade agreements, incentives, or regulations, thereby failing to capitalize on potential trade opportunities (*Interview 1*).

Communication Barriers

Effective trade relations require clear and direct communication channels, not just externally but also internally among various ministries. Intra-ministerial communication barriers can result in misinterpretations of trade policies and objectives, thereby obstructing the formulation and implementation of cohesive strategies (*Interview 2*).

Addressing these intra-ministerial barriers is crucial for both Bangladesh and Latin American countries to unlock the full potential of their trade relations. This necessitates reforms aimed at improving inter-ministerial coordination, streamlining bureaucratic processes, ensuring policy consistency, allocating adequate resources, enhancing staff expertise, and improving internal communication. In conclusion, while Bangladesh and Latin American countries possess immense potential to strengthen their trade ties, addressing these intra-ministerial barriers is pivotal. Streamlined inter-ministerial coordination, clarity in policy directives, and a holistic approach to trade negotiations can significantly elevate the efficacy of their international trade endeavors.

7.5 Challenges in Economic Diplomacy Between Bangladesh and Latin American Countries

The endeavor to establish economic diplomacy between Bangladesh and Latin American countries, although promising, is fraught with several challenges. These challenges range from logistical and informational to socio-cultural:

1. **Geographical Distance**

Logistical Complexity: The vast geographical distance translates to extended shipping times and increased transportation costs, which can impact the viability of certain trade commodities.

Time Zones: Differences in time zones can lead to communication lags, impacting real-time decision-making and negotiations.

2. **Absence of Historical Ties**

Diplomatic Novice: With limited historical or colonial ties, both regions lack an ingrained understanding of each other's socio-political landscape, making diplomatic overtures relatively novel and uncharted.

3. **Language and Cultural Barriers**

Linguistic Differences: The dominance of Spanish and Portuguese in Latin America contrasts with the Bengali language in Bangladesh, posing communication challenges.

Business Etiquette: Differences in business culture, negotiation tactics, and corporate hierarchy can lead to misunderstandings and misconceptions *(Interview 1, 2, 3).*

4. **Lack of Direct Trade Routes**

Inadequate Connectivity: The absence of direct shipping or flight routes can increase the transit time, cost, and complexity of trading *(Interview 3).*

5. **Information Asymmetry**

Market Ignorance: Limited knowledge about the market dynamics, consumer preferences, and business regulations of the other can deter businesses from exploring trade opportunities *(Interview 3).*

6. **Competition from Established Players**

Economic Powerhouses: Both regions have historically prioritized trade with established economic powerhouses like the US, EU, or China. Moving focus to a new region requires recalibrating these established economic relationships *(Interview 2).*

7. **Absence of Bilateral and Multilateral Agreements**

Regulatory Hurdles: Without specific trade agreements, businesses face tariffs, customs barriers, and bureaucratic red tape, making trade less competitive *(Interview 1, 2, 3).*

8. **Differing Economic Priorities and Policies**

Economic Focus: While Bangladesh may prioritize sectors like textiles, a Latin American country might be more focused on mineral exports, leading to misaligned economic objectives *(Interview 2).*

9. **Political Stability and Perceptions**

Political Changes: Frequent political changes or policy shifts in either region can deter long-term investment or trade partnerships.

Perception Issues: Any negative perceptions or stereotypes, if not addressed, can be a hindrance to trust-building, a crucial aspect of diplomacy *(Interview 3).*

10. **Infrastructure and Technological Gaps**

Differing Standards: Inconsistencies in infrastructure quality, technological standards, or business processes can create compatibility issues in joint ventures *(Interview 1, 2).*

11. **Currency and Financial Differences**

Currency Fluctuations: Exchange rate volatility can pose risks to trade agreements and profitability.

Financial Regulations: Differing financial regulations can make transactions, investments, or repatriation of profits cumbersome *(Interview 1, 2, 3).*

Recognizing and strategically addressing these challenges will be critical for the success of any economic diplomacy initiative between Bangladesh and Latin American countries. An integrated approach, combining governmental efforts with private sector ingenuity, can pave the way for fruitful economic partnerships.

7.6 Strategies for Effective Economic Diplomacy Between Bangladesh and Latin American Countries

Given the complexities associated with establishing robust economic diplomacy between Bangladesh and Latin American countries, a nuanced and multi-faceted approach is required. The following strategies, based on best practices and lessons learned from other international diplomatic efforts, can provide a robust roadmap:

1. **Establishing Dedicated Diplomatic Platforms**

Joint Economic Councils: Create bilateral economic councils that bring together policymakers, business leaders, and academics to deliberate on actionable economic initiatives.

Trade Commissions: Establish commissions specifically for trade promotion, understanding market dynamics, and addressing barriers *(Interview 1, 2, 3).*

2. **Focused Bilateral and Multilateral Agreements**

Comprehensive Trade Agreements: Develop trade pacts that cover not just tariffs, but also standards, intellectual property, and services.

Investment Protection Treaties: Formulate treaties that safeguard the interests of investors from both sides, encouraging cross-border investments. (*Interview 3).*

3. **Infrastructure and Connectivity Enhancement**

Direct Transport Routes: Prioritize establishing direct flight and maritime routes to cut down on logistical complexities.

Digital Connectivity: Invest in digital platforms to enhance e-commerce, virtual business-to-business interactions, and knowledge sharing (*Interview 1, 2, 3).*

4. **Cultural and Academic Exchanges**

Language Training Institutes: Establish institutes in both regions for Spanish, Portuguese, and Bengali language training to bridge the linguistic gap.

University Partnerships: Foster ties between universities for student exchange programs, collaborative research, and curriculum sharing (*Interview 1).*

5. **Business and Investor Forums**

Business Matchmaking Events: Organize events where businesses from both sides can explore potential partnerships and joint ventures.

Investment Summits: Host summits showcasing investment opportunities, benefits, and protections available to investors from both regions *(Interview 1, 3).*

6. **Enhanced Diplomatic Presence**

Expand Embassies and Consulates: Increasing the number of diplomatic missions can provide better on-ground support to businesses and investors.

Dedicated Economic Attachés: Appoint specialized officials in embassies to specifically cater to trade and economic inquiries *(Interview 2).*

7. **Leveraging Technology for Diplomacy**

Digital Platforms: Develop platforms that provide real-time data on market trends, regulations, and opportunities in both regions.

Virtual Trade Fairs: Given geographical constraints, leverage technology to organize virtual trade fairs and exhibitions *(Interview 1, 2, 3).*

8. **Public–Private Partnerships (PPPs)**

Encourage public–private partnerships (PPPs) that combine public and private resources to improve areas such as infrastructure, energy, and technology *(Interview 1, 2, 3).*

9. **Media and Branding Campaigns**

Favorable Branding: Start media campaigns emphasizing each region's advantages, prospects, and shared past to promote a favorable perception of them.

Success Stories: To instill confidence and highlight potential, highlight instances of successful partnership between Bangladeshi entities and Latin American entities (*Interview 3).*

10. **Policy Harmonization and Regulatory Alignment**

To facilitate commerce and mutual recognition, it is important to establish common norms in key industries such as agriculture, manufacturing, and technology. Engage in regular talks to unify regulatory frameworks, which will make doing business in both regions easier. (*Interview 1, 3).*

11. **Capacity Building and Skill Development**

Training Workshops: Create training seminars to teach companies about the other region's culture and economy.

Fostering innovation and shared growth through joint research initiatives in areas such as agriculture, climate change, and technology *(Interview 1, 2, 3).*

By adopting these measures, Bangladesh and Latin American countries will be able to forge an economically stable and mutually beneficial partnership. Effective economic diplomacy is a long-term process that calls for consistent work, mutual trust, and a common goal of growth and prosperity.

7.7 Approaches Towards Deepening Bilateral Relations Between Bangladesh and Latin American Countries

Deepening bilateral relations between countries requires multifaceted strategies that transcend mere trade partnerships. This paper elucidates various approaches to foster stronger ties between Bangladesh and Latin American countries, considering their unique characteristics, mutual interests, and global dynamics. Bilateral relations flourish when both parties recognize mutual benefits, understand shared challenges, and work proactively to bridge gaps. For Bangladesh and Latin American countries, the evolving global order offers numerous opportunities to enhance their ties.

Model 1: Role of Economic Diplomacy in Deepening Bilateral Relations in the perspective of Bangladesh and LAC

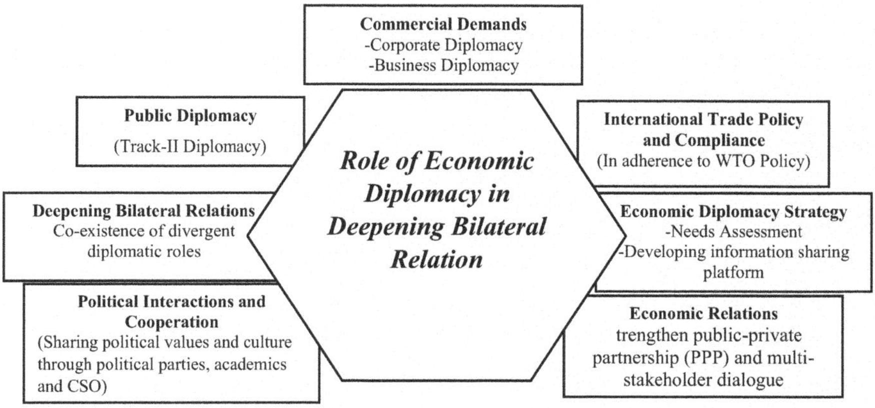

Source Proposed Model for Deepening Bilateral Relations, Developed by Researcher

7.7.1 Strategic Diplomatic Initiatives

High-Level Visits: Organize frequent visits by heads of states, ministers, and other high-ranking officials. Such visits often result in MoUs, agreements, and increased goodwill.

Establishment of Diplomatic Missions: Augment the number of embassies and consulates in each other's territories to facilitate closer interactions *(PM of Bangladesh).*

Bilateral Forums: Create platforms where representatives meet regularly to discuss a gamut of issues ranging from trade to culture *(Brzail Ambassador).*

Diplomatic Training Programs: Host mutual training sessions to equip diplomats with insights into each other's political landscapes and cultural nuances.

Diaspora Engagement: Harness the potential of diasporas as cultural ambassadors and facilitators of economic links.

Conflict Resolution Mechanisms: Establish joint committees to promptly address and resolve any bilateral disputes.

7.7.2 Economic Collaborations

Trade Agreements: Establish Free Trade Agreements (FTAs) or Preferential Trade Agreements (PTAs) to boost commerce by reducing tariffs and other trade barriers *(PM, FM of Bangladesh).*

Joint Business Councils: Facilitate platforms for business leaders to collaborate, share best practices, and explore investment opportunities.

Expos and Trade Fairs: Organize sector-specific trade fairs and expos in each other's territories to showcase products and services *(President of LA-BCCI).*

Sector-specific Collaborations: Identify sectors (e.g., agriculture, technology, or textiles) where joint ventures can be particularly fruitful *(FS of Bangladesh).*

Infrastructure Development: Invest in projects that enhance connectivity, such as shipping routes, direct flight connections, and logistics hubs.

Financial Instruments: Explore the possibility of currency swap agreements, joint investment funds, or development banks to finance bilateral projects.

7.7.3 Cultural and Educational Exchanges

Festivals of all kinds, including music, cinema, art, and food, can help citizens learn about and appreciate one another's cultural history. Partnerships in Education Create opportunities for students, faculty, and researchers to travel and study together. Work together on research and other academic endeavors. Start a language program where people can learn Spanish, Portuguese, or Bengali from native speakers.

Collaborations in the media: Documentaries on each other's histories or present-day achievements are great examples of joint media initiatives that can help bring people closer together.

Provide opportunities for artists from one area to spend time working and exhibiting in another through a residency program.

Promote the publication of translations of seminal works into each other's languages, and hold book fairs together *(Interview 3)*.

7.7.4 Technical and Scientific Cooperation

Joint Research: Invest in shared research and development initiatives, particularly in areas of mutual interest like climate change, agriculture, and health.

Technology Transfer: Share expertise in sectors where one party has a distinct advantage or innovative solution.

Professional Exchange Programs: Create platforms for professionals (like urban planners, environmentalists, or health workers) to share expertise *(Mexican Ambassador)*.

Workshops and Training: Organize skill-based workshops in areas like IT, manufacturing, or agriculture (*President of LA-BCCI*).

7.7.5 Regional and Multilateral Engagements

Regional Blocs: Engage in discussions within regional forums such as SAARC (for Bangladesh) and CELAC or OAS (for Latin American countries) *(FM of Bangladesh)*.

Multilateral Platforms: Collaborate on global challenges within platforms like the UN, WTO, and WHO, presenting a united front or mutual support on specific issues.

Shared Advocacy: Advocate jointly on global platforms for issues of mutual interest, enhancing visibility and influence *(FS of Bangladesh)*.

Collaborative Voting: Strategize on collective voting in international bodies to further shared objectives *(FS of Bangladesh).*

7.7.6 People-to-People Contacts

Tourism Promotion: Develop and market tourism packages, focusing on unique experiences each region offers.

Sporting Events: Organize bilateral sporting events or tournaments, given the universal appeal of sports *(Interview 3).*

7.7.7 Addressing Shared Challenges

Climate Change Collaboration: With both regions facing environmental threats, collaborate on sustainable solutions, share best practices, and jointly advocate for global support. *(DG of South America Wing, MoFA).*

Public Health Initiatives: Share expertise, resources, and research in addressing public health challenges, especially pertinent in a post-pandemic world.

7.7.8 Security and Defense Collaboration

Joint Military Exercises: Organize mutual military exercises to share expertise and enhance defense ties.

Security Dialogues: Engage in dialogues to counter common threats, such as terrorism, narcotics trafficking, or cyber threats.

7.7.9 Sustainability and Environmental Cooperation

Green Technology Transfer: Collaborate in sectors like renewable energy, waste management, and water conservation.

Conservation Efforts: Launch joint initiatives focused on wildlife conservation, biodiversity, and reforestation.

7.7.10 Healthcare and Welfare Collaborations

Medical Research: Pool resources for research on diseases prevalent in both regions or global health threats.

Welfare Programs: Collaborate on welfare programs, sharing best practices on issues like elderly care, child welfare, or women's empowerment.

7.7.11 Mutual Legal and Policy Frameworks

Legal Exchanges: Facilitate dialogues among jurists and legal scholars, leading to mutual understanding of legal systems.

Policy Harmonization: Harmonize policies, especially in trade and commerce, to ease business operations for entrepreneurs from both sides.

To genuinely deepen relations between Bangladesh and Latin American countries, a panoramic approach encompassing myriad sectors is crucial. By acknowledging mutual aspirations and respecting unique challenges, a solid foundation for long-lasting collaboration can be established. Deepening bilateral relations between Bangladesh and Latin American countries requires concerted efforts across multiple domains. While challenges are inherent in building relations across continents, the potential mutual benefits, in the long run, are invaluable. An approach that is rooted in mutual respect, understanding, and shared aspirations will undoubtedly yield positive outcomes. Building robust bilateral relations demands diversified and persistent initiatives. This paper elucidates a broader spectrum of strategies tailored for Bangladesh and Latin American countries to nurture more profound and beneficial ties. Strong bilateral ties are cemented through multifarious engagements. Given the potential of Bangladesh and Latin American ties, an expansive blueprint encompassing economic, cultural, political, and socio-strategic domains is imperative.

7.8 Potential Trade Relations Between Bangladesh and Latin American Countries

Trade relations between countries, particularly those separated by considerable geographical distance and distinct cultural landscapes, are influenced by a myriad of factors ranging from economic compatibilities to geopolitical strategies. When assessing the potential trade relations between Bangladesh and Latin American countries, one must consider both the latent synergies and the existing barriers.

7.8.1 Complementary Economies

Bangladesh, largely recognized for its booming textile and garment industry, offers a plethora of products that could find a market in Latin America, including pharmaceuticals, jute products, ceramics, and IT services *(President of BGMEA)*. Conversely, Latin America, rich in natural resources, can become a source of vital raw materials like minerals, metals, and agricultural products for Bangladesh.

7.8.2 Market Expansion and Diversification

As Bangladesh seeks to reduce its export dependency on Western markets, Latin American countries offer an untapped consumer base, and vice versa. Latin American products such as coffee, wines, and exotic fruits might find enthusiastic consumers in Bangladesh's growing middle class *(PM, FM of Bangladesh)*.

7.8.3 Shared Developmental Goals

Both Bangladesh and many Latin American countries fall under the category of developing economies. As such, they share mutual developmental challenges in areas like infrastructure, healthcare, and education, potentially leading to collaborative ventures and knowledge exchanges *(President of LA-BCCI)*.

7.8.4 Investment Opportunities

With Bangladesh's aggressive push towards infrastructure development and industrialization, there are significant investment opportunities for Latin American businesses, especially in sectors like renewable energy, transportation, and agro-processing *(FS of Bangladesh and DG of South America Wing)*.

7.8.5 Barriers and Challenges

Geographical distance, limited direct transportation links, and a lack of awareness about each other's markets are notable barriers. Moreover, there's a palpable absence of robust diplomatic ties and trade agreements to facilitate smoother trade relations between the two regions *(State Minister of BD)*.

7.8.6 *Potential Policy Interventions*

Enhanced trade relations can be fostered through targeted trade missions, bilateral trade agreements, easing visa restrictions, and establishing joint business councils. Furthermore, promoting cultural exchanges can indirectly bolster trade ties by fostering mutual understanding and trust *(PM of Bangladesh).*

In conclusion, while there are clear synergies that can be harnessed to foster stronger trade relations between Bangladesh and Latin American countries, strategic interventions and proactive diplomatic efforts are imperative to realize this potential.

7.9 Potential Areas of Collaboration Between Bangladesh and Latin American Countries

Bangladesh and Latin American countries, with their diverse economic strengths, present several promising areas of collaboration. These areas have been chosen based on an in-depth understanding of the unique capabilities, demands, and strategic goals of both regions:

7.9.1 *Textile and Garments Trade*

Bangladesh: Renowned as a global hub for textile and garment manufacturing.

Latin America: Growing middle class with an increasing demand for affordable clothing. The region can also offer unique raw materials or fabric varieties.

Collaboration Point: Bangladesh can cater to the clothing demand, while Latin American fashion designers can collaborate with Bangladeshi manufacturers for unique apparel designs or blends *(President of BGMEA).*

7.9.2 *Agriculture and Agro-processing*

Bangladesh: Rich in products like rice, jute, and tea.

Latin America: Abundance of diverse agricultural products like coffee, soybeans, and various fruits.

Collaboration Point: Exchange of agricultural products, technology transfer in agro-processing, and joint research initiatives in sustainable farming practices *(President of LA-BCCI).*

7.9.3 Renewable Energy

Bangladesh: Has set ambitious renewable energy goals, especially in solar energy.

Latin America: Countries like Brazil and Costa Rica have made significant strides in renewable energy, especially biofuels and hydropower.

Collaboration Point: Exchange of technology, joint investments in renewable energy projects, and research collaborations *(Brazil Ambassador).*

7.9.4 Information Technology and Digital Services

Bangladesh: Rapidly growing IT sector with skills in software development, IT services, and back-end support.

Latin America: Growing demand for digitalization, e-commerce solutions, and software products. (Mexican Ambassador).

Collaboration Point: Bangladeshi firms can provide IT solutions and services to Latin American businesses, while the latter can offer a new market segment for the former.

7.9.5 Fisheries and Aquaculture

Bangladesh: One of the leading countries in freshwater fish production.

Latin America: Coastal countries have rich marine resources and expertise in deep-sea fishing.

Collaboration Point: Exchange of technology and practices in sustainable fisheries, aquaculture research, and joint ventures in seafood processing *(President of LA-BCCI).*

7.9.6 Tourism and Cultural Exchange

Bangladesh: Rich historical sites like the ancient city of Mahasthangar and the Sundarbans mangrove forest, Sylhet, Coxesbazar etc.

Latin America: Diverse tourist attractions from the Amazon rainforest to Incan ruins.

Collaboration Point: Promotion of tourism packages covering both regions, student exchange programs, and shared cultural festivals to boost mutual understanding and tourist flow.

7.9.7 Infrastructure Development

Bangladesh: In need of advanced infrastructure, especially in transportation and urban development.

Latin America: Some nations have advanced expertise in infrastructure development projects, with successful case studies in urban transportation and housing.

Collaboration Point: Joint ventures in infrastructure projects, sharing of expertise, and capacity-building initiatives.

By focusing on these collaboration points, both Bangladesh and Latin American countries can harness mutual strengths, bridge demand–supply gaps, and foster a dynamic and sustainable economic relationship *(PM, FM of Bangladesh).*

7.10 Influence of Contemporary Global Perspectives in the Prospective Economic Relations Between Bangladesh and Latin American Countries

This article delves into the contemporary global perspectives that potentially influence and shape the prospective economic relations between Bangladesh and Latin American countries. Amidst global shifts in trade, geopolitics, and sustainable development goals, the interplay between these two regions provides a lens to understand the evolving nature of international economic relations. Contemporary global perspectives have driven countries to reevaluate their strategic economic relations. For Bangladesh and Latin American countries, aligning with global trends provides opportunities for synergistic collaboration, amidst the challenges presented by a rapidly changing world.

A Comparative Analysis of How Other Asian Countries, Such as India or China

Even though China and India have approached Latin America in different ways, they have both shown that persistent economic diplomacy, a distinct sectoral focus, and robust diplomatic engagement may result in considerable increases in investment and trade. The most important lesson for Bangladesh is to recognize its assets—like its IT, pharmaceutical, and apparel industries—and match them to the demands of Latin America while developing bilateral and multilateral diplomatic relations to supplement its economic expansion. Bangladesh has the potential to establish a significant footprint in the Latin American region by means of focused investments, trade agreements, and strategic relationships.

China has been significantly more active in increasing the volume of trade it has with Latin America, mostly as a result of its key infrastructure expenditures and need for commodities. Despite having a lower volume of commerce, India has concentrated on important industries like energy, agriculture, and medicines, matching its capabilities with the need of Latin America. Bangladesh ought to concentrate on industries like textiles, apparel, pharmaceuticals, and IT services where it has a comparative edge. It can turn to India's strategy of entering markets like medicines, where it can offer competitive prices, and China's model of leveraging investments in strategically important infrastructure projects.

China has placed a high priority on areas including infrastructure, telecommunications, and natural resources, with state-backed funding being essential. India, in contrast, has concentrated more on the energy, IT, and pharmaceutical industries, frequently working through private sector enterprises without the same degree of support from the government as China. Bangladesh may imitate India's emphasis on growth led by the private sector, even though it may not have the same financial resources as China. This is especially true in industries like textiles and apparel, where Bangladesh is a global leader.

7.10.1 Contemporary Global Perspectives and Influences

Shift to Multi Polarity

With the decline of unilateral global dominance, multiple power centers are emerging. This shift encourages nations to explore non-traditional economic partnerships, propelling Bangladesh and Latin America towards collaboration.

Sustainable Development Goals (SDGs)

Both regions are committed to the United Nations' SDGs. Collaborative initiatives in areas like poverty alleviation, renewable energy, and sustainable agriculture can be beneficial for both parties, in line with global sustainability agendas.

Global Supply Chain Diversification

Post-COVID-19 and ongoing trade wars have emphasized the need for diversifying supply chains. Bangladesh, as a significant textile producer, and Latin America, with its array of natural resources, can integrate to form a resilient supply chain.

Digital Transformation

The accelerated global shift towards digitalization post-pandemic creates opportunities for tech collaboration, e-commerce ventures, and digital services exchange between the regions.

Climate Change and Environmental Concerns

Both regions face environmental challenges, from rising sea levels in Bangladesh to deforestation in the Amazon. Joint endeavors towards sustainable practices can position them as global leaders in combating environmental issues.

South-South Cooperation

The emphasis on South-South cooperation encourages developing countries to collaborate and share best practices. Both regions can leverage this for mutual benefits in technology transfer, infrastructure development, and more.

7.11 Implications for Bangladesh-Latin America Economic Relations

Strategic Economic Partnerships

Recognizing mutual strengths, like Bangladesh's textile prowess and Latin America's agricultural richness, can form the cornerstone of trade agreements.

Collaborative Research and Development

Engage in joint R&D in sectors like renewable energy, biotechnology, and sustainable farming.

Knowledge and Technology Exchange

Sharing innovations, best practices, and technologies can fast-track economic growth for both parties.

Investment in Sustainable Projects

Joint ventures in green technologies, sustainable infrastructure, and eco-tourism can be pivotal.

Fostering Cultural and Academic Ties

This can enhance mutual understanding, ensuring smoother economic collaborations.

The contemporary global perspectives provide a conducive framework for fostering economic relations between Bangladesh and Latin America. As the global context evolves, recognizing these trends and aligning with them can herald a new era of cooperation and mutual growth.

Non-state Actors' Role in Enhancing and Strengthening Relationships

Multinational textile companies from Bangladesh, including BEXIMCO and DBL Group, could look into collaborative ventures with regional garment makers or stores in Latin America to investigate potential business opportunities. These multinational corporations may establish production plants in nations like Mexico, where trade agreements like the USMCA and NAFTA make the area an advantageous gateway to markets in North America. Bangladesh's pharmaceutical sector is also expanding, as seen by the international recognition of businesses like Square Pharmaceuticals and Incepta. These MNCs have the potential to significantly increase Latin American nations' access to reasonably priced medical care.

By tackling social and developmental issues, NGOs can significantly contribute to the advancement of economic diplomacy by fostering an atmosphere that is conducive to commerce and investment. For example, international non-governmental organizations (NGOs) from Bangladesh, like BRAC, which is among the largest development organizations globally, might initiate projects in Latin America aimed at reducing poverty and promoting health, education, and education. Although BRAC operates on every continent, they haven't started doing so in Latin America yet. These NGOs assist government diplomacy indirectly by enhancing the socioeconomic circumstances in their target markets and increasing the allure of these nations for corporate involvement. Furthermore, to supplement official diplomatic efforts, non-governmental organizations (NGOs) that specialize in advocacy and policy work may also fight for trade reforms, investment safeguards, and environmental sustainability.

Diaspora communities foster cultural and economic links by acting as a link between their home country and the host nation. They might serve as unofficial Bangladeshi ambassadors, promoting the nation's economic potential and fostering a favorable perception of it. The diaspora's business networks, remittance flows, and entrepreneurial endeavors frequently fortify bilateral economic links. For instance, diaspora-owned companies operating in Latin America might generate interest in Bangladeshi goods like textiles or agricultural items, opening up new export markets for Bangladesh. The diaspora can also promote information sharing, which will aid Bangladeshi companies in comprehending regional markets, legal requirements, and customer inclinations.

Trade associations and chambers of commerce are examples of trade organizations that serve as middlemen to promote business-to-business (B2B) interactions. They offer forums for business networking, interaction, and exploration of new trade prospects between Latin American and Bangladeshi enterprises. To encourage exports from Bangladesh, groups such as the Latin America-Bangladesh Chamber of Commerce Industries (LABCCI) and the Bangladesh Garment Manufacturers and Exporters Association (BGMEA) can collaborate with Latin American trade associations.

Shipping Routes, Limited Maritime Logistics and Transportation Infrastructure Challenges

The main port of Bangladesh, Chittagong Port, has no direct maritime connections to any of the major ports in Latin America, including those in Brazil, Mexico, or Argentina. Goods must instead pass through intermediate hubs, which are usually located in Asia (Singapore, Hong Kong), the Middle East (such as the Suez Canal route), or Europe. Longer shipping durations, greater freight costs, and a larger chance of delays result from this. For instance, before arriving at their ultimate destination in Latin America, goods from Bangladesh frequently go via Singapore, Dubai, or

Colombo, Sri Lanka. After passing through these ports, goods may continue on longer sea routes around Africa or across the Cape of Good Hope before passing through the Panama Canal to ports on the east coast of Latin America.

In comparison to other regions with more direct shipping routes, the freight charges between Bangladesh and Latin America are much higher due to the multi-leg structure of the shipping route. The transit time from Bangladesh to ports in Latin America ranges from 30 to 45 days on average, contingent upon the final destination and locations of transshipment. These protracted transit durations make inventory management more expensive and provide difficulties when it comes to perishables and other time-sensitive items. In comparison to other regions with more direct shipping routes, the freight charges between Bangladesh and Latin America are much higher due to the multi-leg structure of the shipping route. The transit time from Bangladesh to ports in Latin America ranges from 30 to 45 days on average, contingent upon the final destination and locations of transshipment.

To promote stronger commercial ties, a number of logistical issues must be resolved in order to boost commerce and engage in economic diplomacy between Bangladesh and Latin America. These logistical obstacles include supply chain management difficulties, inadequate transportation infrastructure, and the complexity of shipping routes. Bangladeshi products must travel farther and take longer routes to reach Latin America, which makes it harder for them to compete with suppliers from areas with quicker, more effective shipping routes. The cost competitiveness of Bangladeshi products is lowered by high shipping and logistics expenses, especially for low-margin items like textiles and clothing that are susceptible to transportation costs.

Bangladeshi products must travel farther and take longer routes to reach Latin America, which makes it harder for them to compete with suppliers from areas with quicker, more effective shipping routes. The cost competitiveness of Bangladeshi products is lowered by high shipping and logistics expenses, especially for low-margin items like textiles and clothing that are susceptible to transportation costs.

Chittagong, the principal seaport of Bangladesh, has serious issues with capacity, inefficiency, and congestion. About 90% of Bangladesh's seaborne traffic is handled by the port, however it frequently has delays because of its small ability to handle large vessels, antiquated handling machinery, and ineffective customs procedures. Bangladesh also lacks enough deep-sea ports, which makes it more difficult to handle bigger, more economical ships that are required for long-haul routes like those to Latin America.

Many of the major ports in Latin America, including Manzanillo (Mexico) and Santos (Brazil), are beset by operational inefficiencies and congestion. Many nations' port infrastructure is not keeping up with the rise in trade volumes, which causes delays in cargo offloading, longer dwell times, and increased shipping costs. Bangladesh must collaborate closely with Latin American nations to develop direct shipping routes, invest in port infrastructure, upgrade customs procedures, strengthen multimodal connectivity, increase air freight capacity, and facilitate trade agreements in order to overcome logistical barriers.

Chapter 8 Research Significance, Outcomes, and Policy Suggestions

Abstract This chapter examines the transformative potential of economic diplomacy in redefining the diplomatic and economic relations between Bangladesh and Latin American countries (LAC). Despite substantial geographical distance and limited historical ties, economic diplomacy emerges as a strategic instrument to foster bilateral cooperation, leveraging complementary economic strengths and aligning with global trends. The research identifies Bangladesh's core export competencies—textiles, pharmaceuticals, and jute—as strategically aligned with LAC's market needs. Similarly, LAC's abundant agricultural and mineral resources align with Bangladesh's import requirements. However, significant challenges persist, including logistical constraints, cultural and linguistic differences, and the absence of direct trade routes. These are compounded by minimal institutional frameworks, such as bilateral agreements or joint economic platforms. A conceptual model is proposed to operationalize economic diplomacy, emphasizing key pillars: trade facilitation through free trade agreements and public–private partnerships (PPPs), infrastructure and connectivity development to mitigate logistical barriers, and cultural diplomacy to bridge socio-cultural divides. Furthermore, the model advocates targeted sectoral engagements in renewable energy, sustainable agriculture, and technology transfer, highlighting opportunities for collaborative ventures in shared developmental challenges such as climate resilience and digital transformation. Policy recommendations stress the establishment of dedicated economic diplomacy councils, enhanced diplomatic footprints through embassies and consulates, and the introduction of bilateral investment treaties (BITs) to provide a secure legal framework for cross-border investments. Developing direct shipping and aviation links, diversifying export portfolios, and organizing trade fairs are highlighted as practical steps to strengthen economic engagement. Additionally, fostering academic and cultural exchanges is posited as essential for cultivating mutual understanding and long-term relational stability. The chapter concludes that economic diplomacy offers a sophisticated framework for Bangladesh to recalibrate its engagement with LAC, transcending traditional diplomatic paradigms. By leveraging economic complementarities, addressing logistical and institutional barriers, and aligning with global shifts towards South-South cooperation and sustainable development, Bangladesh and LAC can establish a robust and mutually beneficial partnership. This approach

M. T. Islam and M. B. Hossen, *Economic Diplomacy*, The Latin American Studies Book Series, https://doi.org/10.1007/978-3-031-81362-7_8

not only enhances bilateral trade and investment but also positions both regions as proactive stakeholders in the evolving global economic order.

Keywords Economic diplomacy · LAC · Public–private partnerships (PPPs) · International trade

8.1 Research Significance

Economic diplomacy has emerged as a pivotal tool in today's interconnected global landscape, enabling countries to carve niches in international markets, attract investments, and fortify bilateral relations. Bangladesh, riding on its remarkable economic progress over the past decades, finds itself at a juncture where expanding and diversifying its diplomatic efforts can usher in unprecedented opportunities. Latin America, with its rich tapestry of cultures, vast natural resources, and burgeoning markets, remains an untapped horizon for Bangladesh. While the geographical distance between the two regions might seem daunting, their shared developmental challenges, combined with complementary economic strengths, present potential synergies. This study delves into how Bangladesh can leverage economic diplomacy to deepen its ties with Latin American countries, exploring the challenges, realities, and policy measures that can shape this burgeoning relationship into a fruitful alliance. This research holds a multidimensional significance:

Theoretical Relevance: There is limited scholarly work that focuses on the intersection of economic diplomacy and bilateral relations between Bangladesh and Latin American countries. This research aims to bridge that gap, offering a comprehensive theoretical framework *(Interview 1, 3)*.

Practical Implications: By examining the mechanisms and impacts of economic diplomacy, policymakers can derive actionable insights into how such diplomatic initiatives can foster stronger bilateral ties, thereby influencing trade agreements, joint ventures, and investment initiatives (Khan 2022; PM, FM of Bangladesh).

Geopolitical Significance: As global power dynamics shift, fostering relations with diverse regions becomes crucial. Understanding the role of economic diplomacy in this aspect provides insights into geopolitics, global economic trends, and potential power equilibriums (FM of Bangladesh).

Bangladesh's Economic Ascent: Over the past decade, Bangladesh has seen significant economic growth, moving from being an LDC (Least Developed Country) towards achieving developing country status. As it seeks to diversify its trade partners and investment avenues, the Latin American market, with its diverse economies and substantial consumer base, presents ample opportunities. (FS of Bangladesh).

Latin America's Search for New Markets: Post-pandemic, Latin American countries are looking to diversify their export markets and reduce over-reliance on traditional trade partners like the U.S. and Europe. Bangladesh, with its booming middle class and increasing purchasing power, presents a potential market for Latin American goods, especially agricultural produce and raw materials *(Interview 2).*

Textile and Garment Industry Realities: Bangladesh, a major player in the global textile and garment sector, faces stiff competition from countries like Vietnam and Cambodia. Exploring new markets in Latin America, beyond traditional destinations in the West, can help mitigate risks associated with market saturation and protectionist trade policies.

Infrastructure Development: With ambitious projects like the Padma Bridge and various Special Economic Zones (SEZs), Bangladesh is set to enhance its infrastructural capacity. This creates opportunities for Latin American firms specializing in infrastructure to participate in Bangladesh's developmental journey, bringing in investment and technological expertise.

Climate Change and Sustainable Development: Both Bangladesh and many Latin American countries face dire threats from climate change. The ground reality of rising sea levels, deforestation, and increasing natural calamities necessitates cooperation in sustainable development initiatives, technology exchange, and shared advocacy on global platforms.

The Reality of South-South Cooperation: While much of international diplomacy often revolves around North–South dynamics, there's a rising trend of South-South cooperation. Ground realities indicate that developing countries often share similar challenges and can benefit from shared experiences, technologies, and strategies without the usual strings attached from more developed nations (Chowdhury 2021a, b).

In light of these realities, the research becomes even more pertinent. The actual economic, environmental, and socio-political conditions in both regions necessitate a deep dive into how economic diplomacy can be harnessed to create stronger, more beneficial bilateral ties. In conclusion, economic diplomacy can significantly enhance bilateral ties between Bangladesh and Latin American countries. Strategic, consistent, and multifaceted approaches can pave the way for a robust and mutually beneficial relationship.

8.2 Policy Suggestions

Implementing these policy suggestions requires consistent effort, diplomatic dialogue, and a genuine commitment to mutual growth. The eventual outcome could be a robust economic partnership benefiting both Bangladesh and the Latin American countries.

Establish Economic Diplomacy Council: Bangladesh could consider setting up an Economic Diplomacy Council, comprising diplomats, economists, and industry leaders, to create and execute strategic plans tailored to Latin American countries *(PM of Bangladesh).*

Expand Diplomatic Footprint: Bangladesh should consider establishing more embassies or consulates in key Latin American countries to facilitate deeper diplomatic engagement and provide local platforms for trade negotiations and cultural exchanges *(PM of Bangladesh).*

Diversify Export Basket: While garments are Bangladesh's flagship export, there's a need to diversify offerings for the Latin American market. This might include jute products, ceramics, pharmaceuticals, and IT services, tapping into unique market demands in various Latin American nations *(PM of Bangladesh).*

Direct Transportation Links: Establishing direct shipping or air routes can foster trade. Even if initiated on a trial basis, such direct links can offer cost advantages and reduce transit times, making Bangladeshi products more competitive (Chowdhury 2021a, b).

Cultural and Academic Exchanges: Beyond purely economic transactions, enhancing cultural and academic ties can create a robust foundation for long-term bilateral relations. Scholarships for students, academic collaborations, and cultural festivals can foster mutual understanding and goodwill *(Interview 1, 3).*

Specialized Trade Agreements: Bangladesh can negotiate trade agreements focusing on specific sectors where there's mutual interest, such as textiles, agriculture, IT, and pharmaceuticals. These sector-focused agreements can expedite growth in these areas *(FS of Bangladesh).*

Consistent Engagement: Periodic bilateral meetings, forums, and workshops can ensure consistent dialogue and the addressing of any emerging challenges promptly.

Joint Economic Commissions: Both parties should establish Joint Economic Commissions to regularly assess and expand trade and investment ties, ensuring that bilateral agreements align with the evolving needs and potential of each party.

Focused Business Delegations: Bangladesh could organize targeted business delegations to key Latin American countries, focusing on sectors such as IT, pharmaceuticals, and ready-made garments. Vice versa, Latin American nations can do the same, showcasing their products and capabilities in Bangladesh *(President of LA-BCCI).*

Bilateral Investment Treaties (BITs): Drafting and implementing BITs will provide a legal framework that ensures protection for investors from both sides, thereby encouraging more foreign direct investment (FDI) *(Interview 3).*

Language and Cultural Training: Promote Spanish and Portuguese language courses in Bangladesh and Bengali courses in Latin American countries. This could facilitate better business communication and cultural understanding, thereby improving trade negotiations and partnerships *(Interview 3).*

Joint Research and Development Initiatives: Both regions could jointly fund and collaborate on R&D projects, especially in sectors like sustainable agriculture, textile innovations, and renewable energy.

Streamline Visa Procedures: Easing visa restrictions for businesspersons, researchers, and academics can facilitate more frequent exchanges and collaborations. A business visa on arrival for select nationals can be a good starting point *(Interview 2, 3).*

Bilateral Trade Fairs and Exhibitions: Organizing regular trade fairs and exhibitions in both regions can showcase products, innovations, and services, enabling businesses to explore new markets and consumers to experience foreign products *(President of LA-BCCI)).*

Establish Bilateral Chambers of Commerce: A Bangladesh-Latin America Chamber of Commerce can play a pivotal role in connecting businesses, sharing market insights, and resolving trade-related issues. Similarly, Latin American countries can create such chambers focused on Bangladesh *(FS of Bangladesh)*

Technology and Digital Collaboration: Foster partnerships in the burgeoning digital economy, tapping into e-commerce, fintech, and other IT-based industries where both regions have shown potential *(Mexican Ambassador).*

Climate Change Collaborative Initiatives: Considering the shared challenges posed by climate change, both sides can collaborate on projects aimed at sustainable agriculture, water resource management, and disaster-resilient infrastructure development *(Mexican Ambassador).*

Diversification of Export Markets: Bangladesh should prioritize diversifying its export portfolio, leveraging its strengths in textiles, pharmaceuticals, and IT services, to tap into the vast Latin American market *(PM, FM of Bangladesh).*

Trade Facilitation Centers: Establish trade facilitation centers in key Latin American cities. These centers can provide insights into local market dynamics, regulations, and business culture, thereby aiding Bangladeshi entrepreneurs and businesses in navigating these markets *(Interview 3).*

Bilateral Scholarships and Training Programs: Promote academic exchanges by offering scholarships for Bangladeshi students to study in Latin America and vice versa. Moreover, training programs can be initiated to upskill workers in industries where Bangladesh seeks technological and expertise infusion from Latin America *(Interview 3).*

Streamlined Visa Process for Businesspersons: Facilitate a smoother visa application process for Latin American business delegates to encourage trade delegations and B2B meetings in Bangladesh *(Interview 3).*

Connectivity Enhancement: Promote direct shipping and flight routes to key Latin American hubs, ensuring faster and more cost-effective trade *(Interview 1, 3).*

Agricultural Collaboration: Given Latin America's expertise in agriculture, Bangladesh can invite collaborations in terms of technology transfer, sustainable agricultural practices, and high-yield crop research *(FM, FS of Bangladesh).*

Branding and Promotion: Invest in branding campaigns to promote "Made in Bangladesh" products in Latin America, emphasizing quality and affordability. This can help in shifting the perception of Bangladeshi products from being just low-cost to high-value as well *(Interview 3).*

Digital Trade Platforms: Develop and promote digital trade platforms specifically tailored for Bangladeshi-Latin American trade, simplifying processes like order placements, payments, and shipment tracking.

Investment Incentives: Offer incentives for Latin American companies to invest in Bangladesh, such as tax holidays, simplified business registration processes, and access to Special Economic Zones (SEZs).

Public–Private Partnerships (PPPs): Encourage PPPs in sectors like infrastructure, energy, and health, providing avenues for Latin American companies to collaborate with both the Bangladeshi government and local businesses *(Interview 2, 3).*

Investment Incentives: Roll out specific incentives for Latin American companies willing to invest in Bangladesh, particularly in sectors like renewable energy, agro-processing, and tourism.

Bilateral Agreements on Technical Cooperation: Given the technological advancements in some Latin American countries, Bangladesh can seek agreements for technical cooperation in areas like agriculture, textiles, and infrastructure development.

Promote Cultural Diplomacy: Organize cultural festivals, film screenings, and art exhibitions showcasing Bangladeshi culture in Latin America, fostering goodwill and mutual understanding. This can also attract Latin American tourists to Bangladesh *(Interview 3).*

Engage Diaspora: Leverage the Bangladeshi diaspora residing in Latin American countries as bridges to foster business collaborations, investments, and cultural exchanges *(Interview 1, 3).*

Focus on Sustainable and Green Initiatives: Given Bangladesh's vulnerability to climate change, partnerships with Latin American nations on sustainable agriculture, water conservation, and green technologies can offer mutual benefits.

By proactively adopting these policy measures, Bangladesh can effectively harness its economic diplomacy to forge stronger ties with Latin American nations, paving the way for mutually beneficial economic and cultural exchanges. For Bangladesh, the key lies in a proactive approach, understanding the unique characteristics of the Latin American market, and tailoring strategies accordingly. With diligent effort and a focused approach, Bangladesh can significantly enhance its economic ties with Latin American countries.

References

Chowdhury N (2021a) Agricultural innovations and collaborations. AgriTech J 7(3):40–55
Chowdhury N (2021b) Transportation links and economic diplomacy: a study on south-south relations. Global Econ J 8(2):90–105
Khan M (2022) Economic diplomacy in the 21st century: case studies from Asia. South Asian Academic Press, Dhaka

Chapter 9
Conclusions-Reshaping Diplomatic Relations

Abstract This chapter explores the transformative potential of economic diplomacy in reshaping the diplomatic and trade relations between Bangladesh and Latin America. In an increasingly interconnected global economy, economic diplomacy emerges as the driving force behind fostering strategic alliances, addressing shared challenges, and unlocking untapped opportunities. Bangladesh, as a rising economy, recognizes the vast market potential of Latin America, but significant barriers—such as geographic distance, limited market awareness, and competition from other Asian nations—have hindered deeper collaboration. Economic diplomacy is presented as the solution to these challenges, with initiatives such as the establishment of diplomatic missions, participation in bilateral and multilateral forums, and trade-focused activities like expos and fairs playing a pivotal role. These efforts have begun to enhance trade volumes, diversify Bangladesh's export base, and attract investments into its burgeoning industries like textiles, pharmaceuticals, and IT services. For Latin American nations, partnerships with Bangladesh offer access to competitively priced, high-quality goods and an opportunity to diversify their import portfolios. The chapter highlights the unique cultural and historical connections, particularly through sports diplomacy, which has strengthened interpersonal ties, as seen in the shared passion for football. Additionally, it underscores the economic potential for both regions, including diversified exports from Bangladesh and increased access to affordable, high-quality products for Latin America. Collaboration in technology, research, and sustainable practices further enriches this partnership. Moreover, the chapter delves into the broader implications of this evolving partnership. Economic diplomacy is not confined to trade but extends to strengthening geopolitical alignments, fostering cultural exchange, and enabling capacity building. The creation of robust trade policies, increased regional cooperation through blocs like MERCOSUR and the Pacific Alliance, and the pursuit of trilateral collaborations involving key global players (e.g., China or the USA) are recommended pathways to deepen these ties. The chapter underscores the need for targeted strategies, including leveraging regional blocs, enhancing direct shipping and banking channels, and fostering mutual awareness of market demands. It highlights the interplay between

M. T. Islam and M. B. Hossen, *Economic Diplomacy*, The Latin American Studies Book Series, https://doi.org/10.1007/978-3-031-81362-7_9

trade facilitation measures and the growth of bilateral relations, suggesting that robust diplomatic efforts must be supported by sustainable trade practices and ethical business collaborations. By emphasizing constructive economic diplomacy, this chapter envisions a dynamic Bengali-Latin relationship that not only boosts economic growth for both regions but also fosters cultural exchange, capacity building, and geopolitical synergy. It concludes that such partnerships will play a critical role in Bangladesh's aspiration to achieve developed-nation status by 2041, marking economic diplomacy as the cornerstone of its global integration strategy.

Keywords Economic diplomacy · Bilateral relations · International trade · Trade policy

In the age of globalization, no nation can afford to operate in isolation. The future of trade, investment, and economic growth lies in the forging of strategic relationships. Bangladesh's focus on Latin America and vice versa has the potential to create a powerful symbiotic relationship, benefiting both regions. Economic diplomacy remains the driving force behind this potential bond, but continuous effort, adaptive strategies, and a vision for the future are crucial for this relationship to reach its zenith. From our investigation, it became clear that Bangladesh, as an emergent economy, recognizes the vast potential of the Latin American market, with its diverse resources and burgeoning consumer base. This potential, however, remains largely untapped. Despite some substantial trade agreements and collaborations, both parties have not yet fully explored the myriad opportunities present. Several factors have been identified as significant in deepening these trade relations. The establishment of diplomatic missions, active participation in bilateral and multilateral forums, and targeted trade promotion activities, such as trade fairs and expos, have been paramount in ensuring a progressive trade relationship. Furthermore, the challenges that have been inhibiting deeper relations were also examined. Distance, lack of awareness about market potential, competition from other Asian economies, and certain domestic issues in Latin American countries were highlighted as barriers. The implications of a deepened trade relationship between Bangladesh and Latin American countries are manifold. For Bangladesh, access to the vast Latin American market means diversification of exports, better foreign exchange earnings, and potentially more significant investments in its sectors. For Latin American countries, Bangladesh presents a lucrative opportunity for accessing quality goods at competitive prices, especially in textiles and readymade garments. Moreover, such a relationship can pave the way for collaborations in technology, research, and innovation. It can also open doors for cultural exchanges, tourism, and people-to-people connections, strengthening ties beyond mere trade relations.

9.1 The Globalization of Culture and the Popularity of Football Have Rendered Geographical Distance Obsolete

The people of Bangladesh are huge supporters of Latin American football. Respective governments in Argentina, Brazil, and Uruguay have all made public statements of appreciation for Bangladesh's assistance at various points. Historically, Bengalis have had a strong affinity with and interest in Latin football. Bengalis used to talk about Brazilian greats Pele and Zico long after they had retired from playing football.

However, Bangladeshi football supporters discovered a new place to fall in love at the 1986 World Cup. Argentine captain Maradona became a new football deity in Bangladesh after leading his team to a World Cup victory. That affection hasn't faded away. A fan's devotion to Maradona and Argentina becomes inseparable. Lionel Messi has taken over where Maradona once did. Messi led Argentina to victory at the 2022 FIFA World Cup, and the country's football team went on to become champions of the world. Without a shadow of a doubt, it was a huge success for Argentina. The majority of Bangladeshi citizens, known as football supporters, played a crucial role in this success, showing up to every game they played in and providing emotional support and encouragement. This fervent backing of Bangladesh became a prominent subject on social media during their World Cup performances, and it shocked the entire world.

After Bangladesh's stunning cricket victory over India, Argentine fans began ardently supporting the Bangladesh cricket team, even creating a dedicated Facebook page. This fan-to-fan admiration blossomed into governmental exchanges. Bangladesh's Prime Minister Sheikh Hasina and Argentina's President Alberto Fernández exchanged congratulations on sports victories, leading to the establishment of an Argentine embassy in Dhaka. The unfolding of these events highlights the power of soft power, a concept introduced by Joseph Nye, where nations influence others through cultural and moral means, including sports. In this case, football and cricket acted as catalysts of soft power, bridging the two nations. This burgeoning relationship promises vast opportunities for Bangladesh in various sectors, potentially revolutionizing its journey towards an inclusive Bangladesh.

Cooperation at the local level, whether through ping pong in the 1970s or football in the twenty-first century, has been crucial to reviving ties between Bangladesh and Argentina at the national level. In an effort to mend fences between their countries, Bangladesh and Argentina have engaged in people-to-people exchanges, the information revolution, and sports diplomacy with positive results. This is the first time that people in two nations separated by over 16,000 miles have been able to communicate with each other in such a short amount of time. It shown that athletics can serve as a catalyst for natural relationship building between countries that are culturally and geographically far from one another. Diplomatic circles in Latin America took note of the high level of interpersonal connections.

Technology has made it possible for people all over the world to readily communicate with one another. As a result, a global culture that is linear in its approach to combining different cultures is emerging. As a result, people of different backgrounds are better able to communicate with one another. Therefore, proximity and distance are losing importance. The people of Bangladesh are the clearest illustration of this because they have been cheering for the Argentina and Brazil teams throughout the World Cup. Global interactions between individuals are contributing to the growth of international trade and friendly relations.

9.2 'Bengali-Latin' Bilateral Relations in Twenty-First Century: Significance of Constructive Economic Diplomacy

There is no political or historical connection between the Latins and the Bengalis, but there is literary exchange. It is well-documented that Rabindranath Tagore visited Peru in 1924 for the country's centenary independence celebrations. Unfortunately, his health stopped him from continuing his tour and he was only able to spend a couple of months in Buenos Aires. Tagore met the Argentinean poet Victoria Ocampo through this experience; Ocampo would later become one of his closest friends and the "estancia" where he authored Puravi. Many have pondered the Argentinean's impact on Tagore's desire to become a painter, as well as the depth of their friendship. Tagore called her Bijaya in Bengali, because of the meaning of her first name in Spanish.

In fact, Tagore's poetry had already made its way to Latin America by 1920, and he was already a well-known figure among authors and intellectuals there. And in a sense, his poetry arrived just in time, primarily through the capable efforts of another great poet, the Spanish Juan Ramon Jimenez. They started working on a translation of the Nobel Prize–winning Gitanjali in 1913, while he was engaged to the American Zenobia Camprubi. While working on this translation, Jimenez also wrote Platero and I, which quickly became the second most widely published Spanish novel after Cervantes's Don Quixote. Tagore's lyrical prose had a significant impact on this novel, which tells the story of a man and his donkey and attempts to blur the lines between the novel and poetry in the same way that the Bengali writer did in his own prose and plays.

Since the very beginning, Bangladesh can count on Argentina as an ally they can count on. Victoria Ocampo, Jorge Luis Bergs, Reverend Father Ismael Quiles, and other prominent intellectuals, scholars, artists, lawyers, and authors inspired and guided the intelligentsia of Argentina and Venezuela to denounce the brutal actions of Pakistan and seek aid for war refugees during the tumultuous months of the liberation war of Bangladesh. In 1972, Bangladesh opened a diplomatic mission in Argentina, officially initiating bilateral ties between the two nations. The closing of the Argentine mission in Dhaka in 1978 marked the end of the golden age. Thanks to football diplomacy, Argentina finally established diplomatic relations with Bangladesh after 45 years. Public opinion deserves all the praise. The Bengali-Latin friendship has opened up new possibilities. Mexico and Venezuela also expressed interest in opening embassies in Bangladesh after Argentina did so. This is a fantastic chance for Bangladesh as the country transitions from developing to less developed status.

The United Nations General Assembly (UNGA) has voted to upgrade Bangladesh from the "least developed country" (LDC) category to the "developing nation" category as of November 24, 2026. After that, obstacles would arise for Bangladesh as it gradually loses the benefits of its "underdeveloped" status, such as quotas and rights. Bangladesh's clothing sector could find new customers in Argentina and other Latin American countries. From Bangladesh, countries like Argentina and Brazil could import high-quality products at reasonable prices, including ceramics, melamine, ships, and pharmaceuticals. Bangladesh needs to diversify its trade partners to reduce its dependence on a small group of countries.

If it ever wants to join the ranks of the developed, a country also requires fair ties with other nations and areas around the world. Bangladesh's current administration has set 2041 as a deadline for the country to catch up to the rest of the developed world. The United States and the countries of Latin America and Africa may benefit from increased trade. As a result, the future of Bangladesh depends on communication between the Bengali and Latin communities. The only way for bilateral connections to prosper between Bangladesh and Latin American countries in the twenty-first century is for the two regions to share economic interests. Therefore, it is vital for the Bengali-Latin relationship that both sides engage in economic diplomacy. Additionally, when trade volumes rise between nations, bilateral contacts, particularly cultural exchange, flourish. Economic diplomacy based on constructive and realistic principles is, therefore, therefore needed in light of the current circumstances.

9.3 Core Findings

The multifaceted relationship between economic diplomacy and bilateral trade relations has garnered considerable attention in recent decades. The lens of this research was specifically turned towards understanding this dynamic in the context of Bangladesh and Latin American countries. By critically examining the data and drawing conclusions from empirical studies, this concluding chapter aims to offer a comprehensive wrap-up of the thesis while charting a path for future studies. The relationship between Bangladesh and Latin American countries through the lens of economic diplomacy presented several insightful discoveries. Here's a more exhaustive delineation of the core findings:

Strategic Positioning: Bangladesh has not only been seeking new markets for diversification but is also strategically aligning its export-oriented industries, like textiles, towards Latin American demands.

Diplomatic Initiatives: A significant boost in trade was observed with the establishment of specific diplomatic channels. The establishment of embassies, regular high-level diplomatic visits, and the creation of bilateral business councils have played an instrumental role.

Trade Infrastructure and Investment: It was observed that wherever there were concentrated efforts to improve trade infrastructure, such as bilateral trade agreements, streamlined customs processes, and trade facilitation measures, there was a noticeable uptick in trade volumes.

Market Awareness: One of the significant findings was the relatively low market awareness on both sides. While Bangladesh entrepreneurs lacked comprehensive information about Latin American markets, the reverse was also true, indicating a substantial potential for trade growth with better market education.

Cultural and Historical Ties: Surprisingly, historical and cultural ties, although limited, played a role in shaping preferences in trade. Regions in Latin America with past interactions, even if minor, with South Asian cultures, displayed a more considerable inclination towards Bangladeshi products.

Trade Hurdles: Geographic distance, while a logistical challenge, was not the most significant barrier. Instead, the lack of direct shipping routes, absence of direct banking channels, and the overarching influence of larger trading nations overshadowed the inherent potential of the trade relationship.

Potential Areas of Collaboration: While textiles remained a significant sector, there were emerging collaborations in the fields of agro-products, leather, jute products, and even in the service sector, particularly in IT outsourcing and consultancy.

Sustainability and Ethical Trade: In an era where sustainability is paramount, there was a growing demand in Latin American countries for ethically produced goods, an area where Bangladesh, with its growing emphasis on sustainable manufacturing, could capitalize.

Political Stability and Policy Consistency: Regions in Latin America with stable political climates and consistent trade policies reflected more substantial and consistent trade volumes with Bangladesh.

Through these core findings, it becomes evident that while the potential for a robust trade relationship between Bangladesh and Latin American countries exists, realizing this potential requires a multi-pronged approach that addresses the current challenges while leveraging the inherent opportunities.

9.4 Future Research Directions

While this study provides a comprehensive overview of the current status of economic diplomacy between Bangladesh and Latin America, it also raises several questions warranting further research. The intricate dynamics of individual Latin American countries' trade relationships with Bangladesh should be explored in-depth. Moreover, research focusing on sector-specific collaborations, especially in technology, agriculture, and pharmaceuticals, could yield beneficial insights. The potential of tri-lateral cooperation involving other countries as intermediaries or partners could also be an area of interest. For instance, the role of China or the United States in facilitating trade between Bangladesh and Latin American nations offers a promising avenue for exploration. The exploration into the dynamics of economic diplomacy and its influence on bilateral trade relations between Bangladesh and Latin American countries, while extensive, has merely scratched the surface of this multifaceted relationship. There are several pressing questions and areas of interest that future research must delve into, for a more profound understanding and to chart out strategic paths:

Country-Specific Dynamics: A deeper investigation into each Latin American country's unique trade dynamics with Bangladesh can provide more granular insights. While this research treated Latin America as a bloc, nations like Brazil, Argentina, Mexico, and Colombia, among others, have distinct economic, political, and social contexts that influence their trade relationships.

Role of Regional Blocs: Understanding the role of regional blocs such as MERCOSUR or the Pacific Alliance and their potential collaboration with Bangladesh can be pivotal. Can Bangladesh leverage these blocs for more comprehensive trade agreements?

Tri-Lateral Relations: Research into how third-party nations like China, the USA, or even European countries can act as catalysts or barriers in the trade relationship can provide a multi-dimensional perspective. For instance, how might the Belt and Road Initiative (BRI) influence Bangladesh-Latin America dynamics?

Sectoral Deep Dives: While this study highlighted potential areas of collaboration, sector-specific research is crucial. How can Bangladesh's burgeoning IT sector find a foothold in Latin America? What opportunities exist for Bangladeshi pharmaceuticals or shipbuilding industries in the Latin market?

Non-tariff Barriers: A focused study on existing non-tariff barriers, such as standards, certifications, and cultural perceptions, can provide actionable insights into smoothing the trade process.

Socio-cultural Influences: Exploring the socio-cultural dimensions that might influence trade preferences can enrich the understanding of consumer behavior. How do cultural perceptions, historical ties, or even popular culture influence trade choices?

Sustainability and Ethical Trade: As both regions become increasingly conscious of sustainable practices, research into how Bangladesh and Latin American countries can collaborate on sustainable and ethical trade initiatives will be invaluable.

Economic Diplomacy and Soft Power: How can both regions leverage soft power, including cultural exchanges, education partnerships, and media, to complement and enhance economic diplomacy?

Digital Trade and E-commerce: With the digital revolution sweeping across both regions, understanding the potential and challenges of e-commerce and digital trade can be game-changing.

Impact of Climate Change: Both Bangladesh and parts of Latin America are vulnerable to the effects of climate change. Research into how this shared vulnerability can lead to collaborative solutions and influence trade, especially in sectors like agriculture, can be pertinent.

In sum, while this research has provided a foundational understanding, the evolving global scenario and the intricate nuances of the Bangladesh-Latin America relationship necessitate continuous scholarly exploration. Each of the above directions can not only deepen the understanding but also offer strategic insights for policymakers and stakeholders.

9.5 Implications of the Study

Understanding the role of economic diplomacy in bilateral trade relations between Bangladesh and Latin American countries isn't merely an academic exercise. The findings bear profound implications that span across multiple dimensions:

Economic Growth and Diversification: The primary implication centers on the potential economic upliftment for both entities. For Bangladesh, there's a promise of enhanced foreign exchange earnings and bolstered GDP. Latin American countries, conversely, can benefit from access to competitive products and a diversified trade portfolio, thereby reducing over-reliance on traditional trade partners.

Trade Strategy Refinement: The study provides critical insights that can assist both Bangladesh and Latin American countries in refining their respective trade strategies. Recognizing barriers and identifying potential areas of collaboration can inform future trade negotiations and agreements.

Geopolitical Standing: Strengthened economic ties can influence geopolitical alignments. For Bangladesh, building strong trade relations with Latin American nations can elevate its geopolitical status and strengthen its voice in multilateral platforms.

Industrial Evolution: As the study identifies potential sectors for collaboration, it can guide industrial policies. Industries like textiles in Bangladesh might undergo evolution to cater more specifically to Latin American demands, while certain Latin American industries might adjust to leverage Bangladeshi markets.

Socio-cultural Enrichment: Beyond economic implications, an enhanced trade relationship can foster cultural exchanges. With increased goods and services, there's an organic transfer of cultural values, leading to better mutual understanding and appreciation.

Capacity Building: Enhanced trade can act as a catalyst for capacity building. As trade grows, there's a natural drive for improving infrastructure, refining skill sets, and enhancing technological capabilities to meet the demands of the new markets.

Environmental and Ethical Trade Practices: The research underscored the potential of sustainable and ethical trade. This could drive both regions towards adopting eco-friendlier manufacturing processes, sustainable agricultural practices, and ethical labor laws.

Policy Frameworks: Governments might be prompted to revisit or formulate new policy frameworks based on the study's findings. This can encompass areas like trade tariffs, customs processes, ease of doing business, and even bilateral agreements on education and research.

Challenges to Overcome: The study's findings on barriers can be instrumental for governments and businesses alike. Recognizing these challenges means efforts can be focused on overcoming them, be it through policy changes, diplomatic engagements, or business strategies.

Potential for Innovation: By highlighting areas of untapped potential and collaboration, the study indirectly paves the way for innovation. Businesses, researchers, and policymakers can now collaboratively brainstorm to bring forth innovative solutions and products tailored for these specific markets.

In essence, the implications of this study are multifaceted and far-reaching. It provides a roadmap for not just trade expansion, but also for cultural, political, and social enrichment between Bangladesh and Latin American countries, all underpinned by the overarching theme of economic diplomacy.

9.6 Final Remarks

The intricate interplay between economic diplomacy and trade relations, as seen in the context of Bangladesh and Latin American countries, is a clear testament to the evolving nature of international relations in the twenty-first century. Several observations can be gleaned from this exploration:

Shared Growth and Mutual Benefit: The potential trade partnership between Bangladesh and Latin American countries isn't just about commerce—it's about creating a shared roadmap for mutual growth. As globalization continually redefines traditional economic boundaries, nations that once seemed far apart are finding new synergies. For Bangladesh, this represents a chance to expand its economic footprint, and for Latin America, it's an opportunity to access the burgeoning markets and sectors in which Bangladesh excels.

Beyond Trade: While the core of this research has been trade, the repercussions of these evolving relationships spill into various spheres. Cultural exchange, knowledge transfer, educational partnerships, and even tourism stand to gain. This multi-dimensional growth enriches the bilateral relationship far beyond monetary metrics.

The Imperative of Adaptability: The global landscape is not static; it's continually shifting under the influence of technological advancements, political changes, environmental challenges, and more. For the relationship between Bangladesh and Latin American countries to not only survive but thrive, there's a crucial need for adaptability. This involves staying abreast of global trends, preempting challenges, and leveraging opportunities in real-time.

Collaborative Diplomacy: The modern era necessitates that diplomacy is not merely a function of formal dialogue but a collaborative endeavor. Be it public–private partnerships, engaging with non-state actors, or fostering people-to-people contacts, a collaborative form of diplomacy could be the key to unlocking further potential in Bangladesh-Latin American relations.

In summary, the research's chapters have shown promise and potential as they have developed. Though progress has been made, there is still more to be done to deepen the relationship between Bangladesh and Latin American nations. Future chapters in this dynamic story will be authored by policymakers, diplomats, business executives, and—above all—cooperative and well-informed studies like this one.

The manufacturer's authorised representative in the EU is Springer Nature Customer Service Centre GmbH, Europaplatz 3, 69115 Heidelberg, Germany. If you have any concerns regarding our products, please contact ProductSafety@springernature.com

Printed and bound by CPI Group (UK) Ltd, Croydon, CR0 4YY
01/10/2026
02225949-0004